PLANT SCIENCE RESEARCH AND PRACTICES

BETEL VINE

NATURE'S "GREEN" GOLD

Plant Science Research and Practices

Additional books in this series can be found on Nova's website under the Series tab.

Additional e-books in this series can be found on Nova's website under eBooks tab.

PLANT SCIENCE RESEARCH AND PRACTICES

BETEL VINE

NATURE'S "GREEN" GOLD

JATINDRA NATH BHAKTA

nova science publishers
New York

NOTICE TO THE READER

Library of Congress Cataloging-in-Publication Data

ISBN: 978-1-53611-941-1

Published by Nova Science Publishers, Inc. † New York

CONTENTS

PREFACE

The betel vine (*Piper betel* L.) has immense social, cultural, economic, nutritional, herbal and medicinal importance as well as a folk (Siddha and Ayurvedha) reputation in each and every plethora of human life from the dawn of civilization. It has been recognized as "Green Gold" and "Household Bank" due to it's afore said multipurpose applications and potential cash crop properties. Although, it possesses a vast array of economic, human and environmental significances, unfortunately it is yet neglected by researchers. Surprising that, though cultivation of the betel vine began long years ago and some researches have recently done on it, there is, however, lacking in conceptual and useful substantial and systematic scientific literatures and books, which can provide an overall concept in this respect. Stemming from this insufficiency of original literatures, the idea of the present books has been conceived by author to draw a quality, reliable and conceptual pictures at a glance considering realistic concepts as well as research literatures in different aspects of the betel vine.

The betel vine is an evergreen, perennial, dioecious and shade loving creeper plant that climbs up trees or other supporting materials. Cultivation of betel vine, betel leaf -chewing and -consumptions are very common and popular conventional customs and practices in many tropical and subtropical countries since ancient era. Due to its increasing demand, recently, it has been recognized as one of the important traditional cash crops in order to develop rural economy of the South-east Asian countries, such as, India, Bangladesh, Nepal, Srilanka, Pakistan, Myanmar, Thailand, Indonesia, Malaysia, Vietnam, etc.

Additionally, the betel leaf has also been greatly contributing as a multipurpose herbal medicine to human life since the Vedic era. It serves as antiseptic, carminative, curing cough and cold, asthma, constipation, pains and sores in throat and chest. It cures stomach ache, diarrhoea, cholera, headache, hysteria, itches, ringworm, and tuberculosis, scales and burns, swelling, bruises, respiratory disorders, constipation, boils and gum sores, etc. It has wound healing, digestive, pancreatic lipase stimulant and deworming properties. Recently, studies surprisingly proved that it has the properties of curing dreaded leukemia or killing blood cancer cells.

It can easily be cultivated by marginal farmers in their little land, which help them to earn alternative cash and enable them to meet their day to day livelihood needs. Thus, it is playing an important role in uplifting the socioeconomic status of rural peoples. Considering these profound beneficial properties, therefore, it has been recognized as "Household Bank" and "Green Gold" of the nature.

The present book specially dealt with fourteen chapters - introduction, morphology and anatomy, cultivar and variety, propagation and growth, chemical properties, importance of betel vine, cultivation technology, farm management, pest and disease, material and tool, production and economy of cultivation, integrated betel vine farming and ecological engineering, advancement in cultivation: problem and scope of research, future prospect and index. The end of each chapter contains the references which is convenient to readers for finding and gaining more depth scientific knowledge in its different respective fields.

The book has been written based on practical concepts gained from long time direct association with its cultivation process and keeping in mind the benefits of farmers, agriculturists, teachers, researchers and students. This resource book has also been designed for trainers to prepare their teaching material that would be useful for students. The information compiled in the book has been collected and synthesized from the widest possible range of different recent literatures, and sources developed from the current researches along with self-field experience.

The exceptional characteristics of the book are: (i) information compiled here basically gained from practical field experience, which is unavailable in other similar books, (ii) consideration of current research information, which is helpful to researchers and scientists, (iii) the field information presented here is especially useful for the common farmers in cultivation practice, (iv) incorporation of integrated farming concept with ecological engineering approach, (v) presentation of a current cost-benefit pictures of the betel vine cultivation, which reveals the economic potential of this cash crop, and (vi) the write up is very simple having plenty of related natural pictures and drawn pictorial illustrations for easy understanding of readers. Apart from scantiness of this kind of book, these above unique fascinating criteria of the present book will make it different from other similar books.

Finally, the readers, farmers, agriculturists, teachers, researcher, students and industrialists will be benefited by enriching their knowledge on various aspects and practical applications of the betel vine presented in this book.

I express my heartiest thanks and gratitudes to my parents, all brothers, family members, teachers, colleagues and those are directly and indirectly associated with betel vine cultivation and chosen the betel vine cultivation as a cash crop and earning resource for uplifting the rural economy. I would also like to extend my sincere thanks to all concern professionals, teachers, students, researchers and farmers for their interest and endless endeavours to improve all aspects of the betel vine in order to achieve a future promising industrial shape in this regard by intervening in its unexplored domains.

Dr. J. N. Bhakta

Chapter 1

Conceptual Overview

1. Introduction

The betel vine (*Piper betel* L.) is an evergreen, perennial, dioecious and shade loving creeper that usually climbs up trees or other supporting materials (Figure 1). It is one of the important cash crops having nutraceutical, medicinal (Das et al., 2016; Bhakta et al., 2016) and recreational properties widely grown and cultivated in moist tropical and subtropical regions of Asia and Pacific, primarily for its edible green leaves. The green betel leaves are popular for mastication among the people of India, Bangladesh, Nepal, Srilanka, Pakistan, Myanmar, Indonesia, Malaysia, Thailand, Vietnam, Laos, Kampuchea, etc. (Ranade et al., 2011).

The glossy heart-shaped betel leaf has aesthetic and ritual significances in various social and religious occasions in South and South-East Asia (Figure 2). It is also auspicious to make offering of betel leaf and areca nut (Figure 3b) in the eve of many occasions such as pious ceremonies, festival, worshiping, wedding ceremonies, etc. (Agri Farming, 2015). The betel vine, areca tree, *Areca catechu* (Figure 3a, c) and lime pot are considered as family members with deep respects in the culture of some

South and South East Asian countries, especially in India, Bangladesh, Vietnam, etc.

Betel vine leaves have a strong pungent aromatic flavour and are widely used as a masticatory along with areca nut (Figure 3c), lime and catechu in the form of betel quid (IARC, 2012) (Figure 4). Chewing of betel leaf is an ancient custom or habit of over 2000 years (Rooney, 1993). Therefore, the habit has the antiquity and universal acceptance in South-East Asia. Folktales reflect on various aspects of betel-chewing practices (Figure 4). Several hundred million people today practice the ancient custom of betel-chewing.

Figure 1. Betel vine cultivated within enclosed system.

Figure 2. Picture showing the decorated betel leaf and areca nut used in various social and religious occasions in South and South-East Asia.

Figure 3. (a) Areca nut tree of different ages, (b) Areca nuts and (c) Typical boxes (pots) containing chewing ingredients used in South and South-East Asia.

Figure 4. Careful preparation of betel quid with betel leaf, areca nut, lime and catechu (left panel) for chewing (right panel) habit.

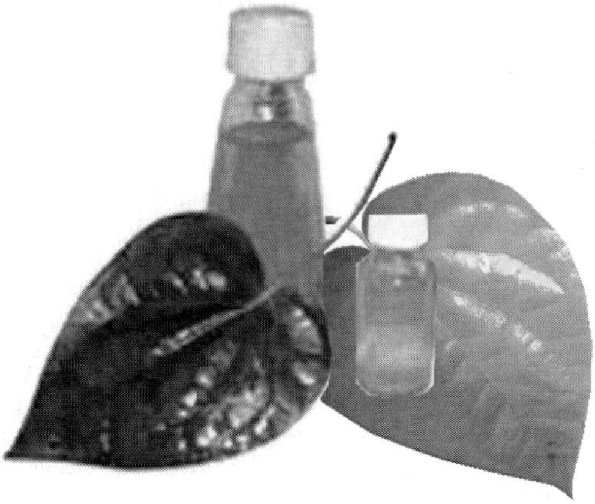

Figure 5. Extracted oil of betel vine leaves used for medicinal purposes.

The betel leaf has much medicinal importance due to having 2 to 3% oil and various active compounds (Figure 5) (Bhakta et al., 2016; Das et al., 2016; Rai et al., 2011; Pradhan et al., 2013; Agri Farming, 2015). It is nutritive and contains vitamin B and C as well as beneficial in facilitating the process of digestion. It moreover serves as good appetizer and helps to remove oral itching and bad odors. Betel leaf contains important components used for producing various medicines in diverse industries (Pradhan et al., 2013; Das et al., 2016). The extracted oil of betel leaves is used as an industrial raw material for manufacturing medicines, perfumes, mouth fresheners, tonics, food additives, etc. It has significant anti carcinogenic properties used for manufacturing of a promising blood cancer drug. Some disputed reports also claim that chewing betel leaves excessively may cause oral cancer. It also acts as an antimicrobial agent by peroxidase, nitric and secretory antibodies (lgA), which offer protection against microbial proliferation in the mouth so that tooth and gum decay is kept under control (Kumar, 1999; Chandra and Sagar, 2004).

There are about more than 150 varieties of the betel vine in the world. Assuming demand, the different variety of the betel vine is widely grown using the tress of agro-forestry or commercially cultivated in organized farm in some areas of above-mentioned countries. The betel vine cultivated as an important commercial crop in the well-organized farm which demands constant care from the farmers, viz. regular watering, manuring or fertilization, disease treatment, crop harvesting and lowering the plants, etc.

In association with betel vine cultivation, many integrated farming processes are practiced in different parts of India, Bangladesh and Nepal. These are fishery, poultry, vegetable cultivation, horticulture and floriculture, etc. This integrated cultivation of the betel vine is designed by following the sustainable approach for the benefit of both artificial ecosystem/environments constructed and human society. It also provides the farmers extra benefit along with the betel vine cultivation. The betel vine can be cultivated using the organic wastes scientifically and removes substantial amounts of greenhouse gas, carbon dioxide from the environment. In these view points, it plays significant roles in recycling the waste materials of the environment for the production of an economically valuable plant biomass. Thus, it can be recognized as an ecological engineering approach for sustainable environment.

On account of the above, it is apparent that the betel vine is one of the most important commercial and cash crops, which plays a vital role in overall livelihood and economic security of substantial and considerable percentage of rural farmers in afore said countries. However, the cultivation of betel leaf is suffering from various problems such as – disease, marketing, high labour cost, elevated cost of materials required for cultivation, application of hazardous and non-eco-friendly chemicals (fertilizer, pesticides, insecticides and antimicrobial agents) and lack of research, especially concerning the advanced cultivation and management technologies. To solve the problems, it can be cultivated using the approach of eco-friendly and ecological engineering technologies. The present book deals with various aspects of the betel vine – basic concept, cultivars, chemical properties, significance, cultivation, management, marketing and cost benefit, advancement in cultivation technology, future prospect, etc.

2. Etymology

In 1550s, the name *Piper* was probably originated from the Sanskrit term *'Pippali'* (meaning long pepper) and the name 'betel' or 'bétele' (Portuguese) might have come from the Indians Malayalam word the 'Vettila' or 'Vetila' or from the Tamil word verrilai. The 'Vetel' in Malayalam means Betel, and the 'ila' means simple leaf (Peter, 2004).

3. Vernacular Name

The deep green heart-shaped leaves of the betel vine are called in various popular names in different parts of the growing countries of the world. The nomenclature of the betel vine is the general craze with growers to name the varieties after their village or town (Pradhan et al., 2013; Agri Farming, 2015). The vernacular names of betel vine are listed in table (Table 1).

In India, the betel vine is called by various names. It is known as Paan in Hindi or Assamese or Oriya or Bengali, Tambula and Nagavalli in Sanskrit, Nagbael in Marathi, Naagarbael in Gujarati, Vetrilai in Tamil, Tamalapaku/Nagballi in Telugu, Veeleyada yele in Kannada and Vettila in Malayalam.

4. History of Origin and Geographical Distribution

The betel-chewing is a deeply rooted ancient custom in the subtropical countries. According to Candolle (1884), *P. betel* might have originated in the Malaya Archipelago. Burkill (1966) described the native place as Central and East Malaysia where the crop was cultivated and spread through tropical Asia and Malaysia.

Table 1. List of vernacular names of betel vine

Vernacular name	Language
Nagavallari, Nagini, Nagavallika,Tambool, Saptashira, Mukhbhushan, Varnalata	Sanskrit
Sirih, Sirih melayu, Sirih cina, Sirih hudang, Sirih carang, Sirih kerakap	Malaysia
Betel, Betel pepper, Betel vine	English
Bulath	Sinhalese
Maluu	Khmer, Cambodian language
Malus	Tetum, Austroasiatic language in Indonesian West Timor
Malu	Tokodede, languages of East Timor
Plū	Mon, Austroasiatic language
Kun	Burmese
Tambol, Tambool	Arabic, Persian
Serasa, Cabe	Semang
Kerekap, Kenayek	Jakun
Jerak	Sakai
Sirih, Suruh, Bodeh	Javanese
Pelu	Thai
Bileiy	Dhivehi, Maldivian
Bulung samat	Kapampangan, languages of the Philippines
Daun sirih	Malaysian, Indonesian
Pupulu	Chamorro, Austronesian
Ikmo	Tagalog, Austronesian language
Pu	Laos
Trầu	Vietnamese
Gaweud/Gawed	Kalinga
Buyo	Bikol, Central Philippine languages

The first domestication areas of the areca palm tree which are located somewhere in Malaysian archipelagoes. On the basis of historical, epigraphic records and philological as well as archaeological evidence, some researchers proposed that original home of the betel vine is to be Indonesian archipelago. Anthropologists suggested that the remains were to be probably areca nuts and trees of betel, have been found in the spirit caves in Northwest Thailand dating back as to 7000 – 5500 BC, which is even before systematic and organized agriculture came to be practiced. [Remains of what C. F. Gorman (1970: 98) suggested to be probably areca nuts, dating to 7000–5500 BC, have been found at the Spirit Cave in northwestern Thailand, but the clue of their domesticity needs to be scientifically confirmed (Gorman, 1970)]. There have been similar findings in Timor in Indonesia going backward to 3000 BC and in the blacked teeth of a human skeleton in Palawan in a Philippines going back to 2600 BC. According to Skeletons bearing evidence of betel-chewing date back to about 3,000 BC and were found in the Duyong Cave in the Philippines.

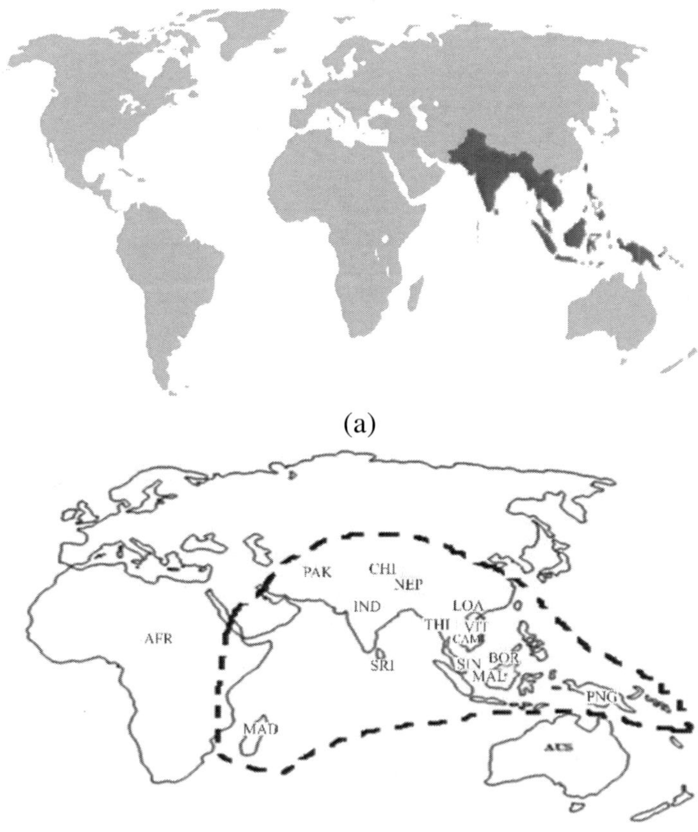

(a)

(b)

Figure 6 (Continued)

(c)

Figure 6. (a) Shaded parts are the regions of betel leaf and areca nut consumption around the world, (b) Region within the dotted lines shows major areas of *Piper betel* consumption [Where, AFR: Africa, MAD: Madagascar, PAK: Pakistan, IND: India, NEP: Nepal, SRI: Sri Lanka, THI: Thailand, CAM: Cambodia, MAL: Malaysia, PNG: Papua New Guinea, BOR: Borneo, SIN: Singapore, LOA: Laos, VIT: Vietnam (Pradhan et al., 2013; Ranade et al., 2011)], (c) Major Betel vine growing areas in India (Pradhan et al., 2013).

However, it is believed that betel vine is native to central and eastern Malaysia and was taken into cultivation more than 2500 years ago throughout Malaysia and tropical Asia. Later on, it reached Madagascar and East Africa much later and was also introduced into the West Indies. Written Chinese sources from the period of the Tang dynasty (AD 618-907) described South-East Asia as a region of betel users. Betel chewing was widespread in South India and South China when the first Europeans arrived in the 15th century. With known ethno medicinal properties, today these plants are distributed in several countries and also cultivated in South and South-East Asian countries, especially in India, Bangladesh, Nepal, Srilanka, Pakistan, Myanmar, Indonesia, Malaysia, Vietnam, Laos, Kampuchea, Thailand, Singapore and the Far-East (Figure 6a,b,c) (Kumar, 1999; Pradhan et al., 2013; Rnade et al., 2011). It has also proposed that betal vine has entered in India during early Gupta's period and slowly merged into Indian culture (Ahuja and Ahuja, 2011).

5. Scientific Name and Taxonomic Classification

The scientific or botanical name of the betel vine is *Piper betel* Linn. It belongs to the family of piperaceae, i.e., the black pepper family. The taxonomic classification of the betel vine is as follows:

Kingdom	:	Plantae
Unranked	:	Angiospermae
Division	:	Magnoliphyta
Class	:	Magnolipsida
Order	:	Piperales
Family	:	Piperaceae
Genus	:	Piper
Species	:	betle
Botanical name	:	*Piper betle* Linn.

6. Habitat and Ecology

The betel vine is wild in nature. It is shade loving perennial evergreen climber of tropical origin. It likes to grow in humid tropical forests based ecological conditions or high shady land. Generally, betel widely grows under shade of various trees in the forest as a wild vine.

The betel vine grows well in loamy or clay loamy soil, which is friable, rich in organic and has the good drainage capacity with pH of about 5.6 to 8.2 (CSIR, 1969; Guha and Jain, 1997; Guha, 2006). It has an ability to grow in the wide range of soils from sandy loam to clayey loam (Agri Farming, 2015). It can also be successfully grown in well-drained and fertile soils in wet to dry climatic zone. The soil with low water permeability is not favourable to its growth due to the problem of water logging. The soil of high electric conductivity (i.e., >1 milli *mohs/cm*) is not appropriate for betel vine cultivation. Especially, the lateritic and clay loam soils are highly suitable for betel cultivation. It can also be grown in sandy, sandy loam, red lateritic, black cotton and coastal alluvial soils. It cannot tolerate the waterlogged, saline and alkali soils. Debanath et al. (1985) reported that integrated nutrient management (INM) was always advantageous from a long-term perspective both in terms of cost of production and soil health (Thomas et al., 2013).

The betel vine thrives best under humid tropical condition with cool shade. It prefers a diffused sunlight, cool humid with considerably high relative humidity and regular supply of moisture to the soil is essential. It grows best under the shaded, tropical forest based ecological conditions with relative humidity and temperature ranging from 40 to 80% and 15 to 40°C respectively. It can tolerate temperature between 10ºC to 40ºC.

Well-distributed annual rainfall enhances the growth of betel vines. It flourishes in areas with an annual rainfall of 225 to 475 cm and is cultivated at altitudes up to 1000 m. However, in the areas with lower rainfall (1500 - 1700 mm) the crop is cultivated with small and frequent irrigations, i.e., generally every day in summer and every 3 to 4 days in winter, whereas adequate water drainage is required during the rainy season (Jana, 1995; Mishra et al., 1997; Guha, 2006).

The betel vine is a sun-loving plant, but it produces better quality leaves in the wet and intermediate zones rather than in the dry zone. Appropriate shade and irrigation are essential for successful cultivation of betel crop. It cannot withstand extremely cold or hot winds flow because it damages the tender tips of vines and causes their wilting. Since, hot dry winds are harmful and retard the growth of the vine, a cool humid and shady atmosphere is artificially created inside the vineyard or Boroj.

In India, the South West coastal regions of Kerala, North Kanara in Karnataka, Plain lands of West Bengal and Norm East hilly regions of Assam provide appropriate conditions for cultivation. Under unsuitable climatic condition, it is cultivated by providing protection on sides and tops for shading and maintaining the suitable conditions of humidity. The betel vines (usually the male plants) are cultivated throughout India except the dry northwestern parts. Further, the female plants also rarely produce any flower or fruit in the Indian climate.

Generally, November to February months are optimal period for establishment of betel vine in India. Because, this period is of low temperature and cool humid, suitable for establishment of the betel vine. The crop requires special care for its cultivation like sterilization of soil, which is achieved mainly by solarization and aeration of soil by tilting several times.

References

Agri Farming (2015). http://agrifarming.in/betel-leaf-farming-information.

Ahuja, S.C, Ahuja, U. (2011). Betel Leaf and Betel Nut in India: History and Uses. Asian Agri-History, 15(1):13

Bhakta, B., Bhakta, U. Ghanti, S., Jana, B. B., Ghanti, P., Yamane, S., Bhakta, J. N. (2016) Betelvine: A green gold for sustainable economic upliftment in rural India. *Int. J. Env. & Tech. Sc.* 3:76-82.

Burkill, I. H. (1966). A dictionary of the economic products of the Malay Peninsula. (pp. 249-260). Malaysia: Ministry of Agriculture and Co-Operatives.

Candolle, A. D. (1884). Origin of cultivated plants, London, pp. 13.

Chandra, G. and Sagar, R. L. (2004). Harvesting green gold: cultivation of betelvine in sundarban. *Indian Farmers Digest*, 37(3): 5-13.

CSIR (Council of Scientific and Industrial Research) (1969). The Wealth of India. 8: 84-94. CSIR, New Delhi, India.

Das, S., Parida, R., Sandeep, I. S., Nayak, S. and Mohanty, S., (2016). Biotechnological intervention in betelvine (*Piper betle* L.): A review on recent advances and future prospects. *Asian Pacific Journal of Tropical Medicine*, 2016: 1–9. http://dx.doi.org/10.1016/j.apjtm.2016.07.029

Debanath, P. K., Sengupta, K., Sengupta, C. and Chatterjee, B. N. (1985). Betelvine cultivation in West Bengal. *Indian Fmg,* 25(7): 22-23.

Gorman, C. F. (1970). Excavations at spirit cave, North Thailand, some interim interpretations. *Asian Perspectives*, 13: 79-107.

Guha, P. (2006). Betel leaf: The neglected green gold of India. *Journal of Human Ecology,* 19(2): 87-93.

Guha, P. and Jain, R. K. (1997). Status report on production, processing and marketing of betel leaf (*Piper betle* L.). Agricultural and Food Engineering Department, IIT, Kharagpur, India.

IARC (International Agency for Research on Cancer) (2012). Personal habits and indoor combustions. WHO Press, 20 Avenue Appia, Geneva, Switzerland

Jana, B. L. (1995). Gram Banglar Arthakari Phasal-Paan (in Bengali). [Betel leaf: A Cash Crop of Villages of Bengal], Calcutta, India.

Kumar, N. (1999) Betelvine (*Piper betle* L.) cultivation: A unique case of plant establishment under anthropogenically regulated miclimatic conditions. *Indian journal of History,* 34(1): 231-239.

Mishra, D., Mishra, S. K., Acharya, A. and Das. J. N. (1997). Annual Report of All India Coordinated Research Project on Betelvine. Department of Plant Pathology, College of Agriculture, Orissa University of Agriculture and Technology, Bhubaneswar, India.

Peter, K. V. (2004). Handbook of herbs and spices. CRC Press LLC, 2000 Corporate Blvd, NW, Boca Raton, FL, USA.

Pradhan, D., Suri, K. A., Pradhan D. K. and Biswasroy, P. (2013). Golden Heart of the Nature: *Piper betle* L. *Journal of Pharmacognosy and Phytochemistry*, 1: 147-167.

Rai, M. P., Thilakchand, K. R., Palatty, P. L., Rao, P., Rao, S., Bhat, H. P. and Baliga, M. S. (2011). *Piper betel* Linn. (Betel vine), the maligned Southeast Asian medicinal plant possesses cancer preventive effects: time to reconsider the wronged opinion. *Asian Pacific J Cancer Prev,* 12: 2149-2156.

Ranade, S. A., Soni, A., and Kumar, N. (2011). SPAR profiles for the assessment of genetic diversity between male and female landraces of the dioecious betelvine plant (*Piper betle* L.). In: Oscar Grillo (Ed.), *Ecosystems Biodiversity,* pp. 443-464.

Rooney, D. F. (1993). Betel Chewing Traditions in South-East Asia. Kuala Lumpur, Oxford University Press.

Thomas, U. C., Chandini, S. and Thomas, A. (2013). Integrated nutrient management studies in betelvine *(Piper betel L.). International Journal of Applied Research and Studies,* 2(7): 1-9.

Chapter 2

MORPHOLOGY AND ANATOMY

1. INTRODUCTION

The betel plant, a branching vine with alternatively arranged heart-shaped leaf, is known as "green heart" of the nature due to shape of its leaf (Figure 7). It can grow and creep on plain soil or climb up the trees or with the support of other materials. The length of the betel vine is not definite, and it depends on the growth rate. Generally, male betel vines grow up to a height of 20 m with a stem diameter of 15 to 20 cm. The mature betel vine consists of five major parts - stem, leaf, root, flower and fruit. The flower and fruit are occasionally found. Despite the applied significance of this valuable crop, very little morphological and anatomical studies of the betel vine have been done (Verma et al., 2013; Balasubrahmanyam et al., 1994; Rawat et al., 1989; Sharma et al., 1983; Sharma et al., 1987; Guha 2006; Lakshmi and Naidu, 2010). Based on research literatures available, the morphological and anatomical structures of the betel vine plant have been described here as follows:

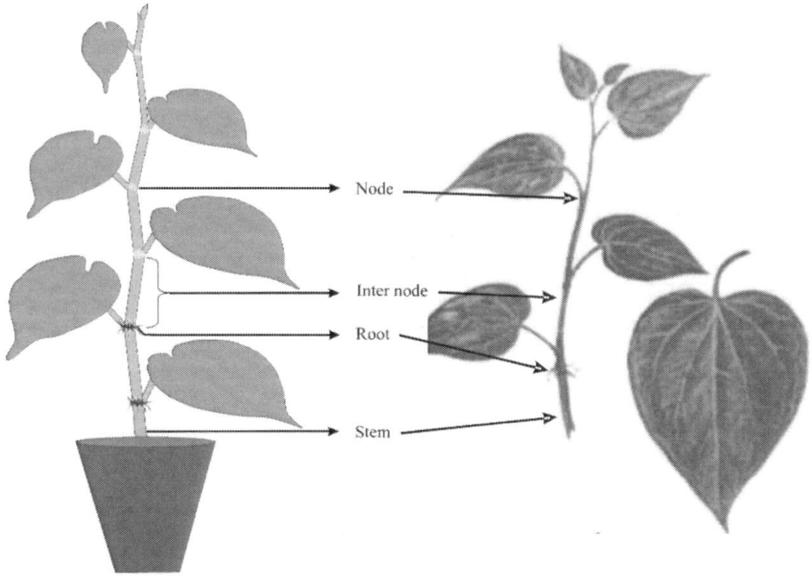

Figure 7. Betel plant and heart-shaped leaves.

2. Morphology

2.1. Stem

The stems are dichotomous, articulate and semi-woody in nature (Figure 8). It is commonly green or pinkish green in colour. It is stout and cylindrical or bilaterally pressed with orthotropic vegetative branches. The stem is with nodes, internodes and branches. The nodes are conspicuously swollen and about 3 mm in diameter, which gives raises the adventitious roots those help to hold the support and apical shoot. The shape of internode is elongated somewhat cylindrical in nature and is about 2.5 to 4 cm long. The branches are sprouted from the nodes which are called as the lateral vine. Every node does not sprout the lateral vine, and it depends upon the species of vine and climatic condition. The branches grow in length throughout the year as that of the main stem.

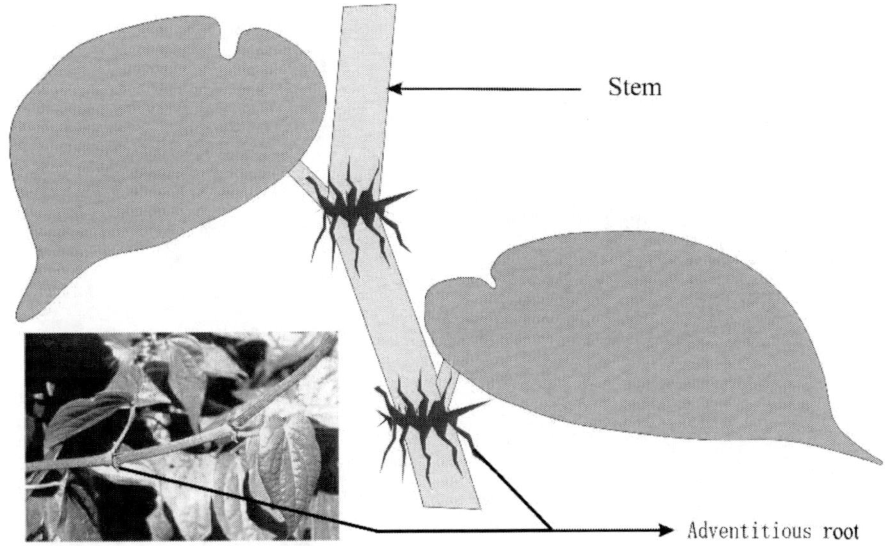

Figure 8. Stem and adventiticious root of the betel vine.

2.2. Leaf

Leaf of the betel vine is the most economically valuable and medicinally important part apart from the other parts of plants. It is characteristically shiny heart shaped with different sizes. It sprouts from the node and alternatively arranged (Figure 8, 9). It is simple yellowish green to dark green in colour with glossy upper surface. The odour of leaf is characteristic and pleasant. The betel leaves are aromatic with varied taste, ranging from sweet to pungent due to the presence of essential oils. The mature betel leaves are simple, stipulate, bifarious, petiolate, thick and often unequal. It consists of two parts – petiole and lamina. The petiole is roughly triangular and outline has deep furrows and ridges. It is about 2 to 15 cm long. The lamina of leaf is smooth with even surface and reticulate venation. The size of a lamina is ranged from about 7 to 18 cm in length and 5 to 14 cm in width and is varied in different cultivars. The leaf is ovate oblong broadly ovate cordate or obliquely elliptic complete glabrous coriaceous 10 to 18 cm long and 5 to 10 cm broad acuminate oblique and rounded base (Periyanayagam et al., 2012; Vasuki et al., 2011; Lakshmi and Naidu, 2010; Pradhan et al., 2013). It has an entire margin and also with an often undulated margin. The apex of leaf is acuminate with often unequal base. The leaf is palmately veined and generally about 5 to 7 veins arise from the bases and reach to the tip. Lakshmi and Naidu (2010) reported comparative leaf morphology of some Indian *Piper betle* L. cultivars (Table 2).

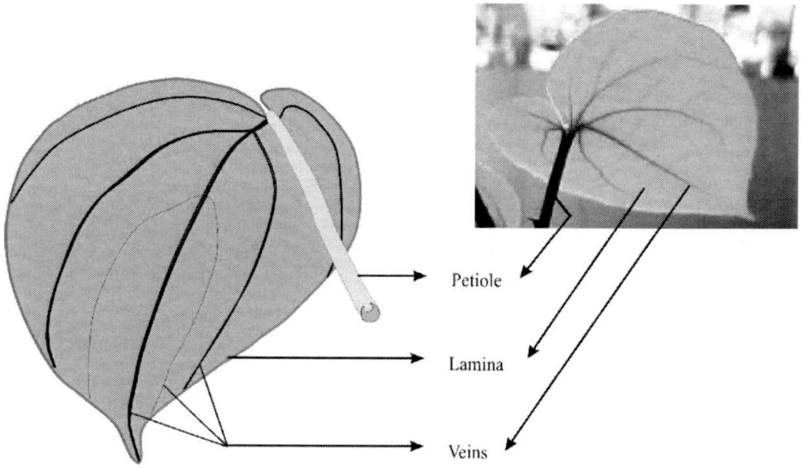

Figure 9. The leaf of betel vine.

Figure 10. Adventitious root of the betel vine firmly attached to support stick.

Table 2. Comparative leaf morphology of *Piper betle* cultivars

S. No.	Name of cultivar	Colour and Texture of leaf	Length (cm)	Width (cm)	Nature of leaf tip	Number of secondary veins
1	*P. battle* L. Cv. Calcutta, Madhya Pradesh	Yellowish green, glossy upper surface and coriacious	8	7.5	Acute	7
2	*P. battle* L. Cv. Desipan, Uttar Pradesh	Dark green, glabrous	7.5	5	Curved acuminate leaf tip	5 Intra marginal veins present asymmetric
3	*P. battle* L. Cv. Desawari, Uttar Pradesh	Greenish yellow, glabrous	8.5	7	Curved acuminate leaf tip	5
4	*P. battle* L. Cv. Ghazipur, West Bengal	Dark green, glossy upper surface and coriacious	14	12	Acute	7
5	*P. battle* L. Cv. Bangladeshi, West Bengal	Dark green, coriacious	13	11	Acute	7
6	*P. battle* L. Cv. Benarasi, Uttar Pradesh	Dark green, glossy upper surface and coriacious	13	11.5	Acute	7
7	*P. battle* L. Cv. Jaleswar, Madhya Pradesh	Yellowish green, glossy upper surface and coriacious	15	14	Acute	7
8	*P. battle* L. Cv. Vishnupuri Pan, Madhya Pradesh	Dark green, coriacious	15	14	Acute	7
9	*P. battle* L. Cv. (Kapoori) Tuni, Andhra Pradesh	Light green, glabrous	12.5	7.5	Acute	7
10	*P. battle* L. Cv. Saunfia Pan, Madhya Pradesh	Dark green, coriacious	9.5	7	Acuminate	7

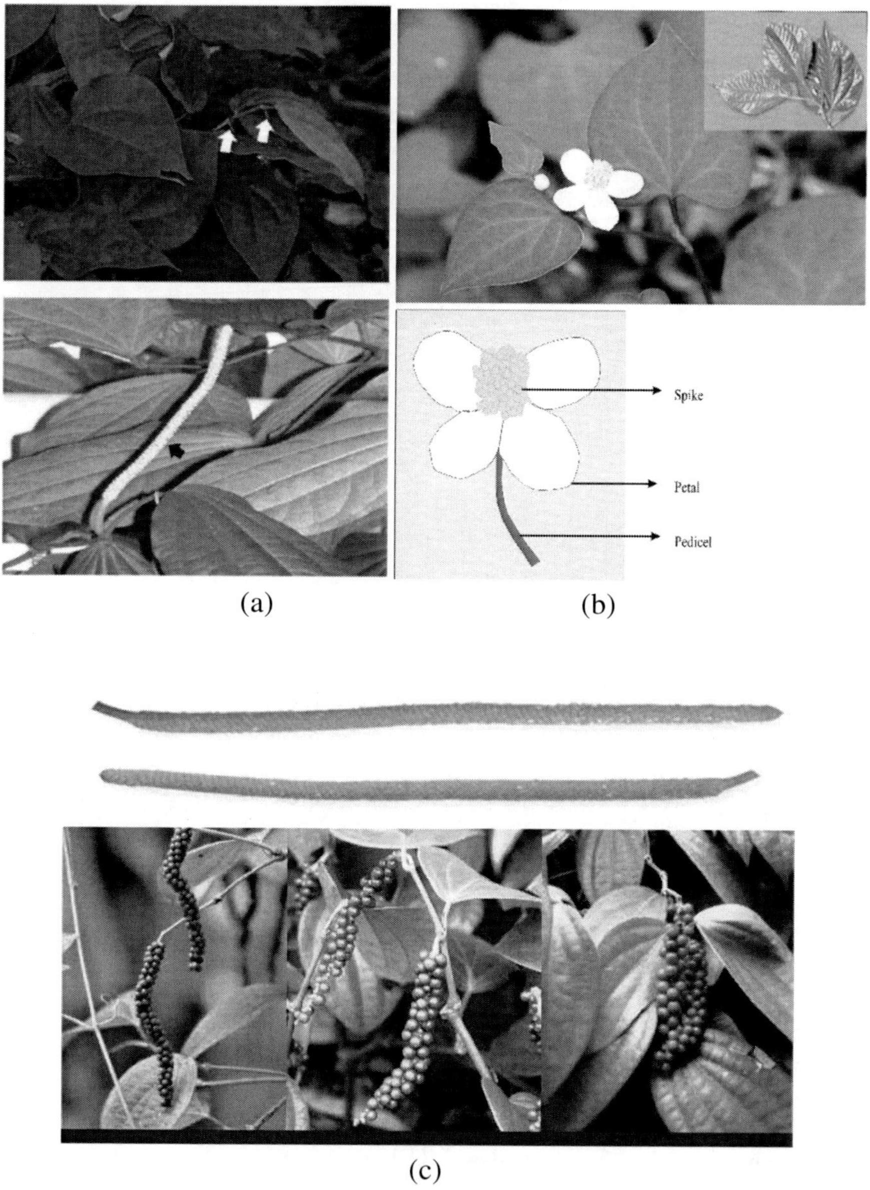

Figure 11. (a) Photos of the female (upper panel) and male (lower panel) betel vines, (b) The solid white arrows (top panel) and the solid black arrow (bottom panel) point to the female and male inflorescences respectively (Ranade et al., 2011), (c) Different stages' fruits of the betel vine.

2.3. Root

The root is adventitious in nature. It is branched threat like or cylindrical in shape and dark brown in colour (Figure 10). The growing end or tip of the root is white in colour which turns to dark brown in mature stage. A number of threats like roots are grown around the nodes. The length of root is varied from 0.5 to 4.0 cm depending on the climatic conditions and species variations. The vine penetrates their roots when they come in contact with support for holding and climbing up.

2.4. Flower and Fruit

The male and female betel vines are found under the suitable environmental conditions. Flowers of plant are found when they are generally 8 to 10 years old. Therefore, flowering is rare, because the plants are replanted usually in every five or six years under cultivation. Except the dry northwestern parts, the betel vines (usually the male plants) are cultivated throughout India. The female plants also rarely produce white flower spikes or fruit in the Indian climatic conditions (Figure 11). The inflorescence is cylindrical, pendulous spike and flowers are naked, unisexual, dioecious, fairly long, peduncled (3-10 cm long) and oppositifolius. Female spikes are about 3.5 to 6 cm long. Male spikes are dense, cylindrical, 8 to 10 cm long, sub-pendulous, consisting of numerous unisexual bracteate flowers.

The fruits are orange in colour, drupaceous, seen very rarely and often sunken in fleshy spike. It is 3 mm in diameter and consists of about 10 to 20 seeds, which are poor in germination. There is much variation in the chromosome numbers reported in the betel vine: figures 2n = 26, 32, 52, 58, 62, 78 and 195 have been given. The most frequent chromosome number is 2n = 78 for the majority of cultivars and varieties (Jose and Sharma, 1984; Pradhan et al., 2013).

3. Anatomy

3.1. Stem

The internodes of adult stem have the arrangement and structure of the vascular bundles. A thick cuticle is present outside the single-layered epidermis. The epidermal cells are papillose. Typical three-celled hydathode with a balloon-like or disc like apical cell and bicellular trichomes (hairs) is commonly found on the surface.

The cortex is distinguishable into 2 to 7 layered collenchyma. There are several layered parenchyma in stem. Sometimes this parenchymatous region shows the lobed outline due to the presence of vascular bundles immediately below. It is distinguished into three more or less distinct regions: (1) outermost region of about three layers of parenchyma, (2) a middle region of 5 to 6 layers of discontinuous chlorenchyma and (3) an innermost region of 4 to 6 layers of parenchyma with some chloroplasts. Secretary cells (oil cells) are also present in the cortex. No clear pericyle or endodermis could be seen. Chibber (1913) described a definite endodermis in very young stems, and he also interpreted the parenchyma cells below the collenchyma region as pericycle (Figure 12).

There is a very conspicuous mucilage canal in the centre of the internode. In addition to this, a ring of smaller canals is also found separating the two rings of vascular bundles.

The numerous vascular bundles which are collateral and open are arranged in two concentric rings. The bundles of the outer most ring are always greater in the number than those of the inner, but they are not uniform in their size. This ring has about 25 bundles. The second ring consists of 10 to 12 bundles.

It is interesting to note that one of the bundles of the inner ring is transversely oriented or twisted through 90°. Chibber (1913) reported several such twisted bundles in the same species.

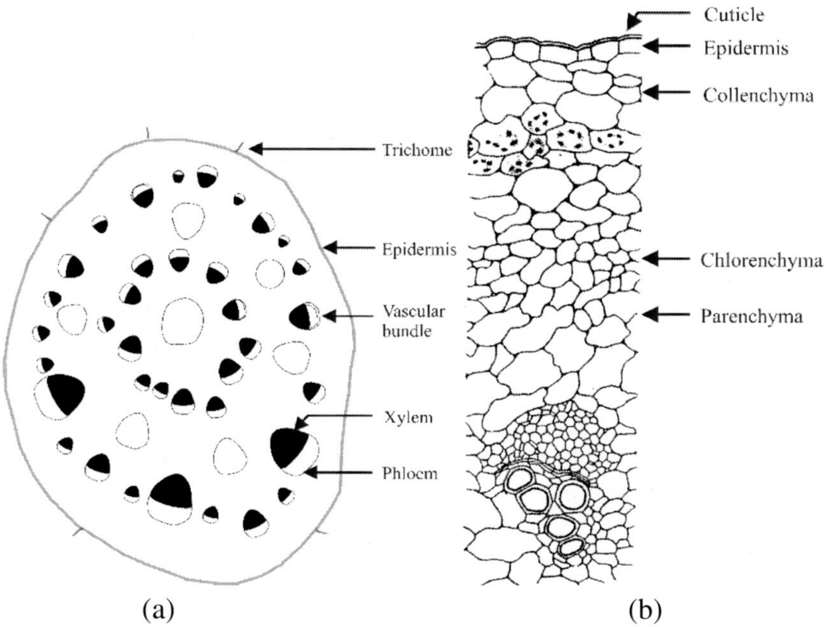

Figure 12. Ccross-sections (a and b) of stem (internode) of the betel vine.

3.2. Leaf

The leaf is dorsiventral and mesomorphic with fairly prominent semicircular midrib of the even outline. It has a multiple epidermises in both the adaxial and abaxial surfaces with glandular, tector trichomes. In almost all the cultivated varieties of *P. betle*, the leaves consist of four layered upper and two layered lower epidermis (Lakshmi and Naidu, 2010).

Midrib – the adaxial side of midrib is flat with circular secretory canal and small secretory cells, while the abaxial side is hemispherical and has tiny secretory cells. The epidermis of midrib is thin and has spindle-shaped cells with smooth surface. Inner to the epidermis is a narrow zone of 2 or 3 layers of collenchyma cells. Vascular bundles are single ovate collateral with cluster of xylem elements and thick arc of phloem (Figure 13).

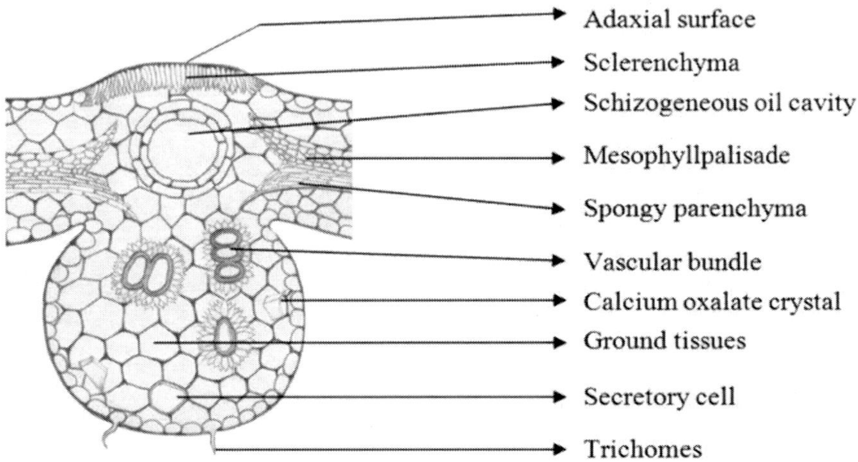

Figure 13. Transverse section of betel leaf with midrib (Pradhan et al., 2013).

Lamina – it is about 160 to 170 μm thickness with adaxial epidermis of 20 μm thickness and abaxial epidermis of 12 μm thickness with short cylindrical glandular trichomes (pearl glands). The cuticle is thick on the upper epidermis and thin on the lower epidermis. The cells of the outer epidermal layers on both sides of the leaf are small, that possess tannins and oils. The sub epidermal cells on the abaxial side are enlarged, and they store water. Crystal and oil reserves are found in the subepidermal cells on both sides. The palisade layeres are well distinguished they are double layered short wide compact cells and mesophyll cells are 3 to 4 layered and small lobed. Thick walled irregular secretory cells are seen with dense contents of probable an essential oil as shown in Figure 13.

The leaves are hypo-stomatic and tetra-cytic stomatal complexes, which is the characteristic feature of the Piperaceae. The abaxial epidermis is stomateferous while adaxial epidermis is apostomatic. They have a single row of short, wide vertically oblong palisade cells and spongy mesophyll of 3 or 4 layers with secretory cells of 25 to 30 μm in diameter (Figure 13). Occasionally stomatal complexes are cyclocytic but twin stomata and just aposed type are also seen. Conjugation tube like structures is present between the twin stomata in Kapoori-Tuni variety. The trichomes are glandular, which have unicellular apical cell and a short pedicel. The pedicle has a thicker wall, surrounded by 5 or 6 epidermal cell arranged as a rosette disc like manner. The apical cell of trichome is slightly pointed or clavate shaped. Presence of twin glandular trichomes is common in Kapoori Tuni variety (Lakshmi and Naidu, 2010).

Petiole – it is composed of epidermis with thick cuticle, sclerenchyma cells, schizogenous oil cavity, vascular bundles and secretory cells (Figure 14). In the cortex, both parenchyma and collenchyma occur in longitudinal strips. Cortical collenchyma is subepidermic. The central cylinder of the petiole composed of scattered vascular bundles of different dimensions and devoid of endodermis and pericycle. The large schizogenous oil cavities are found in the medullary regions. Crystals and oil reserves are observed in the cortical collenchyma and in parenchyma regions of the petiole. The crystals are mostly prismatic types. Tracheoid idioblasts like structures have been noticed from the vessel elements of midrib and petiole. These tracheoids are suggested to store water (Fahn, 1997), and they are connected to the vein endings present in the leaves and cortex in the petiole. These characteristics are typical of xeromorphic anatomy of leaves. The petiole showed numerous covering trichomes (Mubeen et al., 2014).

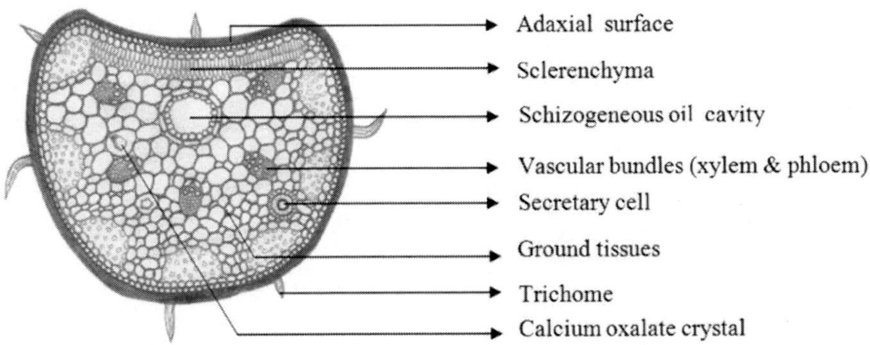

Figure 14. Transverse section of petiole of betel leaf (Pradhan et al., 2013).

REFERENCES

Balasubrahmanyam, V. R., Johri, J. K., Tripathi, R. D., Chaurasia, R.S. and Rawat, A. K. S. (1994). Economic Botany Information Services, National Botanical Research Institute, Lucknow, India, p. 24.

Chibber, H. M. (1913). The morphology and histology of *Piper betel* Linn. *Linn. Soc.* 41: 357-383.

Fahn, A. (1997). Plant Anatomy. 4th Edition. Aditya Books Ltd., New Delhi, India 233-234.

Guha P. 2006. Betel leaf: The neglected green gold of India. *Journal of Human Ecology,* 19(2): 87-93.

Jose, J. and Sharma, A. K. (1984). Chromosome studies in the genus *Piper betel* L. *J Indian Bot Soc.,* 63:313-319.

Lakshmi, S. B., and Naidu, K. C. (2010). Comparative Morphoanatomy of *Piper betle* L. cultivars in India. *Annals of Biological Research*, 1(2): 128-134.

Mubeen, M., Periyanayagam, K. S. and Sathik Basha, S. (2014). Anatomical Investigation on the leaves of *Piper betle* (L) var. Sirugamani 1(SGM1) links an Ethnomedical important Medicinal plant and its Pharmacognostic relevance. *International Journal of Pharm Tech Research,* 6(1): 244-251.

Periyanayagam, K., Jagadeesan, M., Kavimani, S. and Vetriselvan, T. (2012). Pharmacognostical and Phytophysicochemical profile of the leaves of *Piper betle* L. var Pachaikodi (Piperaceae)-Valuable assessment of its quality. *Asian Pacific journal of tropical biomedicine*, 2(2): S506-S510.

Pradhan, D., Suri, K. A., Pradhan D. K. and Biswasroy, P. (2013). Golden Heart of the Nature: *Piper betle* L. *Journal of Pharmacognosy and Phytochemistry*, 1: 147-167.

Ranade, S. A., Soni, A., and Kumar, N. (2011). SPAR profiles for the assessment of genetic diversity between male and female landraces of the dioecious betelvine plant (*Piper betle* L.). In: Oscar Grillo (Ed.), Ecosystems Biodiversity, pp. 443-464.

Rawat, A. K. S., Tripathi, R. D., Khan, A. J. and Balasubrahmanyam, V. R. (1989). Essential oil components as markers for identification of Piper betle L. cultivars. *Biochemical Systematics and Ecology,* 17(1): 35-38.

Sharma, M. L., Rawat, A. K. S. and Balasubrahmanyam, V. R. (1987). Studies on essential oil of betel leaf. *National Botanical Research Institute*, 31(4): 319-324.

Sharma, M. L., Rawat, A. K. S., Balasubrahmanyam, V. R. and Singh, A. (1983). *Indian Perfume,* 27: 91-92 (Chemical Abstracts, 101, 11993).

Vasuki, K., Senthamarai, R., Kirubha, T. S. V., Balasubramanian, P. and Selvadurai, S. (2011). Pharmacognostical studies on leaf of Piper betle. *Der Pharmacia Letter*, 3(5): 232-235.

Verma, A., Kumar, N. and Ranade, S. A. (2004). Genetic diversity amongst landraces of a dioecious vegetatively propagated plant, betelvine (*Piper betle* L.). *Journal of Biosciences,* 29(3): 319-328.

Chapter 3

CULTIVAR AND VARIETY

1. INTRODUCTION

The betel vine is one of the ancient and heritage crops of some countries. Its growers invariably named their cultivars according to local or vernacular names. Therefore, these cultivated different types of betel vines are nothing but cultivars (landraces). Many of these landraces differ from each other in several organoleptic properties. Scrutiny of the landrace names and their etymology suggests that a given landrace may be named differently in different regions and more than one landraces may have the same name. Despite the importance of this crop, very little research has been done so far on genetic variations of several characters of betel vines. Due to the absence of systematic attempts to resolve this problem of nomenclature as well as vegetative propagation of the betel vines, most of these names are as ancient as the cultivation of the betel vine itself. A few individual and isolated efforts have been made to recognize the different landraces and to identify similar or dissimilar types of betel vines by characterization using various morphological, biochemical and genetic techniques.

2. CULTIVAR CHARACTERIZATION

2.1. Morphological Characterization

Some researchers have attempted to differentiate various cultivars on the basis of morphology of betel vines. However, the following morphological properties are considered to characterize and differentiate some betel vine cultivars:

1. Colour of leaf
2. Shape and size of leaf
3. Length and width of leaf
4. Nature of anterior (tip) and posterior parts of leaf
5. Colour, shape and size of petiole of leaf
6. Colour, number and arrangement of primary and secondary veins of leaf
7. Shape, number and index of stomata in leaf

Lakshmi and Naidu (2010) revealed that the cultivars have differences in leaf morphological characteristics such as – colour, texture, length, width, nature of tip and number of secondary veins of leaf as described in the table (Table 3) considering some cultivars.

2.2. Biochemical Characterization

Besides the morphological characters, the varieties or betel vine cultivars are determined by phytochemical characterizations. The analyses of electrophoretic patterns of peroxidase isoenzyme and chromatophoretic patterns of extracted oils of betel vine leaf are generally used as important techniques for determining the cultivars. Dissimilarity in electrophoretic and chromatophoretic patterns identifies the various cultivars of the betel vine.

Table 3. Characteristics of leaf morphology of some Indian *P. betle* cultivars collected from different states of India (Lakshmi and Naidu, 2010)

Name and origin of cultivar	Colour and texture of leaf	Length (cm)	Width (cm)	Nature of leaf tip	Number of secondary veins
Calcutta, Madhya Pradesh	Yellowish green. glossy upper surface and Coriacious	8	7.5	Acute	7
Desi pan, Uttar Pradesh	Dark, green, glabrous	7.5	5	Curved Acuminate	5 intra marginal veins Present, asymmetric
Desawari, Uttar Pradesh	Greenish, yellow, glabrous	8.5	7	Curved Acuminate	5
Ghazipur, West Bengal	Dark green, glossy upper Surface and coriacious	14	12	Acute	7
Bangladeshi, Werst Bengal	Dark green coriacious	13	11	Acute	7
Benarasi Uttar Pradesh	Dark green, Glossy Upper Surface and Coriacious	13	11.5	Acute	7
Jaleswar Madhya Pradesh	Yellowish Green, Glossy Upper Surface and Coriacious	15	14	Acute	7
Vishnupuri pan, Madhya Pradesh	Dark Green, Coriacious	15	14	Acute	7
Kapoori/Tuni Andhra Paradesh	Light green glabrous	12.5	7.5	Acute	7
Saunfia pan Madhya Pradesh	Dark Green Coriacious	9.5	7	Acuminate	7

Based on biochemical and morphological characteristics, the betel vine cultivars are grouped into five distinct cultivars - Bangla, Kapoori, Meetha, Sanchii and Desawari. The taste and pungency of leaf are largely depended on the qualitative and quantitative variations of important phytochemical constituent, essential oils, which are also significantly associated with characterization of various betel vine cultivars. Such as, eugenol ($C_{10}H_{12}O_2$), caryophyllene ($C_{15}H_{24}$), methyl eugenol ($C_{11}H_{14}O_2$) and para cymene ($C_{10}H_{14}$) are commonly present in all types of betel vine cultivars, whereas anethol ($C_{10}H_{12}O$) is only found in Meetha varieties, and eugenol is the major constituent of essential oils in Bangla cultivar. Anethol (32%) and Eugenol (40%) are the chief phytoconstituents in Metha and Bangla varieties, respectively, which are responsible for their specific characters of good test and high market demand. Balasubrahmanyam et al. (1995) reported that landraces with the prefix *Desi* in their names invariably refers to the landrace '*Bangla*' in West Bengal, landraces '*Kapoori*' in Maharashtra and landrace '*Desavari*' in Madhya Pradesh in India. However, cultivar specific biochemical profiling is insufficient so far. Therefore, an extensive research effort should be made to uncover the biochemical differentiation among betel vine cultivars.

2.3. Genetic Characterization

Modern molecular technology is capable of easily isolate deoxyribonucleic acid (DNA) from cell of an organism and to employ for finding the variation as well as for identification of organism at species or strain levels by applying DNA fingerprinting technique. Currently, it is an important tool for identification of species and varieties of organisms. It is obvious that the research work on molecular identification of varieties of the betel vine is insufficient so far. The study of genetic variation among the landraces using molecular or biochemical methods is, however, scanty as well as crop improvement work is also limited to germplasm evaluation and selection (Ravindran, 2000). Rawat and Balasubrahmanyam (1988) recognized five distinct varieties of betel vine landraces from the germplasm that differ in their morphology and chemistry. They are namely, *Bangla*, *Kapoori*, *Meetha*, *Sanchii* and *Desawari*.

Recently, DNA profiling of betel vine using the polymerase chain reaction (PCR) and random amplified polymorphic DNA (RAPD) methods are employing to identify duplicates or sort the germplasm and to estimate genetic diversity among some betel vine cultivars (Ranade et al., 2002; Verma et al., 2004). Using this technique, a few disease-tolerant or resistant varieties were also identified through the screening of cultivars. Using the RADP technique, the study of comparative genetic variation of some Indian betel vine cultivars (Table 4) showed the different DNA band patterns (Figure 15) (Patra et al., 2011). It indicated that the cultivars are genetically different, which is also cleared from their position in the phylogenetic tree (Figure 16) (Patra et al., 2011)

Additionally, the gender distinction and determination have not yet been well studied in the betel vine. The study has employed SCAR (sequence characterized amplified regions) marker in determining the sex type of betel vines and this technique will also help in designing economical breeding programmes (Khadke et al., 2012). It is a PCR and DNA sequence based techniques, which represents a sequence of genomic DNA fragments of cultivars. For genetic characterizations of all betel vine varieties, an extensive study using DNA fingerprinting techniques is important in order to draw a clear picture in this respect.

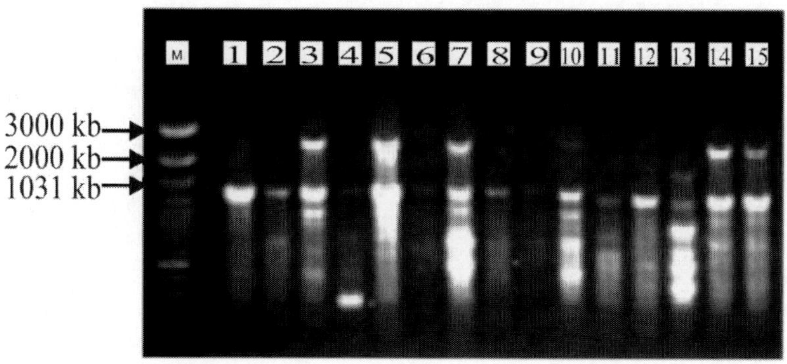

Figure 15. DNA banding patterns of different fifteen cultivars (Patra et al., 2011).

Table 4. Name and source of collection of fifteen Indian cultivars used in the analysis of DNA profiling and phylogenetic analysis (Patra et al., 2011)

Sl. No.	Name of cultivars	Source of collection
1	Godi bangale	Kakatapur, Orissa
2	Nua bangala	Chandanpur, Orissa
3	Balipana	Puri, Orissa
4	Birkoli	Balangir, Orissa
5	Kali mahata	Puri, Orissa
6	SGM-1	Sirugamani, Tamilnadu
7	Kali bangala	Howrah, West Bengal
8	Halisahar sachi	24-Parganas, West Bengal
9	Awani	Jorhat, Assam
10	Gandhi pana	Jorhat, Assam
11	Ramtek bangala	Ramtek, Madhya Pradesh
12	Maghai	Jutalpur, Madhya Pradesh
13	Bangala mandasore chinthalpudi	Chinthalpudi, Andhra Pradesh
14	Karapaku bangala	Baputal, Andhra Pradesh
15	Andaman local	Andaman Nicober

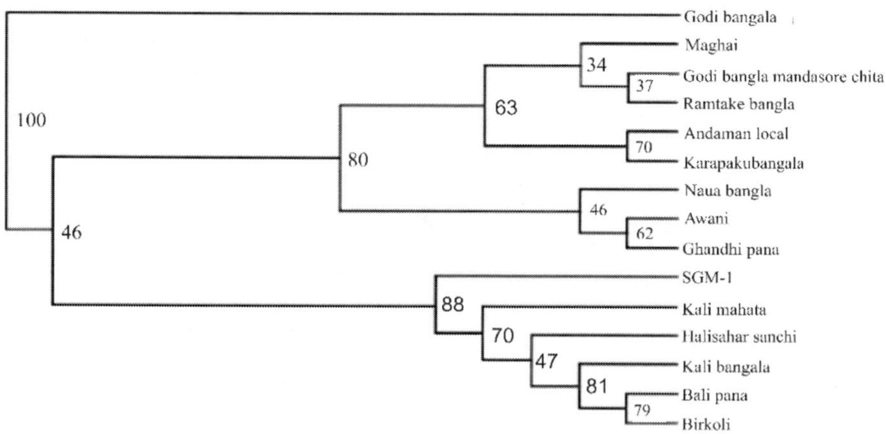

Figure 16. Phylogenetic analysis showed the phylogenetic positions of fifteen different Indian betel vine cultivars (Patra et al., 2011).

3. Varieties and Properties

Several landraces are grown in many countries. Though there are many varieties of betel vines in the world, its exact species number is not clear according to literature available. However, about 125 to 150 cultivars (landraces) of the betel vine are grown by cultivators and recognized by traders in India (Verma et al., 2004; Gamble, 1928; Lakshmi and Naidu, 2010; Patra et al., 2011). According to Guha (1997, 2006), Maity (1989) and Samanta (1994), there are about 100 varieties of the betel vine in the world, of which about 40 are found in India and 30 in West Bengal Some of these are Deshi, vali, nalekatj Kurche, Sanchi, Remtaks, Pangara, Kapufi, Banaresi, Calcutta, Madres, Malpari, Chandan, Kali. Among others "Maneru", "Ratadalu" and "Galdalu" are popular betel vine varieties of high export quality. Karpurakodi, Kallarkodi, Revesi, Karpuri, SGM 1, SGM (BV) - 2, Vellaikodi, Pachaikodi, Sirugamani 1, Anthiyur kodi, Kanyur kodi and Bangla type are popularly cultivated in some parts of India. The most popular local variety Cheelanthikarpuram is largely selected for cultivation, because, its leaves are dark green, broad, and coarse-textured with good storage quality. The most important cultivars commonly grown in India have been listed in the following (Table 5).

Although, there are many varieties, the following six cultivars of betel vine viz. Kappori, Bangla, Meetha, Sanchi and Maghai are mainly cultivated in India systematically for commercial production:

1. *Bangla:* The Bangla cultivar is moderately vigorous. The leaf is green to dark green, cordate in shape, thick with broad lamina, acuminate, coarse to touch and pungent. Petiole or leaf stalk is long, thin and widely spaced due to longer internodes. No or very few lateral vines grown in this cultivar.
2. *Meetha:* The Meetha pan is normally sweet in taste and palatable. The leaf is green, thick, waxy and large in size. It is nonfibrous and remains soft after maturation. Its stalk is small, and less pungent than Bangla variety. This variety is commercially most important

and grown exclusively in West Bengal and monetarily highly rewarding. Lateral vines are found to grow in this variety.

3. *Kappori:* The leaf of kapoori is greenish yellow, ovate in shape, smooth with an acute tip, juicy and non pungent. It consists of 4 to 6 sub veins along with prominent mid vein. It produces a large number of laterals/vine. The yield of this betel vine is higher compared to other cultivars. It is extensively cultivated in south India.

Table 5. Important cultivars of betel vine in India (Ravindran et al., 2000)

Sl. No.	Cultivars	Pungency*
1.	Bangla (Madhyapradesh)	P
2.	Bangla (Uttar Pradesh)	P
3.	Bangla Nagaram (Uttar Pradesh)	P
4.	Calcutta (West Bengal)	P
5.	Calcutta Bengal (West Bengal)	P
6.	Deshi Calcutta (West Bengal)	P
7.	Desvar Mahoba (Uttar Pradesh)	MP
8.	Ghanghatte (West Bengal)	P
9.	Godi Bangla (Orissa)	P
10.	Halisahar Sanchi (West Bengal)	P
11.	Kakir (Bihar)	P
12.	Kalipatti (Maharashtra)	P
13.	Kappori (Bihar)	NP
14.	Kappori (Orissa)	NP
15.	Karapaku (Andra Pradesh)	P
16.	Karpuri (Tamil Nadu)	NP
17.	Kaljedu (Andhra Pradesh)	NP
18.	Maghai (Bihar)	P
19.	Meetha Pan (West Bengal)	Sweet
20.	Nov Bangla (Orissa)	P
21.	Ramtek Bangla (Maharashtra)	P
22.	S. G. M. 1 (Tamil Nadu)	MP
23.	Sachi pan (Assam)	P
24.	Sangli Kapoori	NP
25.	Tellaku (Andhra Pradesh)	NP
26.	Vellai Kodi (Tamil Nadu)	NP

*MP = mildly pungent; P = pungent; NP = non-pungent

4. *Sanchi:* The plant has an oval leaf blade with long tapering apex. Plant is more or less similar to Bangla variety. The leaf is varied from a medium to large in size. It has 6 to 8 sub-veins along with the main vein. Growth of the vine is not so fast. The leaf is with a high pungent aroma. Disease resistance and preservation capacity of sanchi variety are also very high.
5. *Desawari:* The leaf is ovate with a small acute tip. It has a prominent mid vein with 4 to 6 sub veins arranged near the mid veins.
6. *Maghai:* It has smaller leaves than that of other cultivars. Leaf is sweet and palatable. It is famous all over India for high quality leaves.

Table 6. Merits and demerits of different cultivars

Name of Cultivars	Merits	Demerits
Bangla	High growth and yield, high preservation efficiency of leaf (i.e., leaf can be preserved for long period), high percentage of essential oil content, high tolerant to adverse environment, high disease-resistant	Pungent and fibrous
Meetha	Sweet taste, presence of anethole compound, low fibrous	Require extensive rearing and management, low yield, low disease-resistant
Kappori	High leaf number and yield, resistant to some diseases, require less rearing and management,	Low leaf preservation efficiency, low tolerant to adverse environment, low essential oil content
Sanchi	High preservation efficiency of leaf, high disease-resistant	Pungent in taste
Desawari	Good taste, low fibrous	Low disease and cold - tolerant, high disease susceptible, low leaf preservation efficiency
Maghai	High yield, resistant to some diseases	Leaf thick and leathery, pungent in taste

In India, the above six cultivars are the most popular, and their leaves are used both for domestic consumption and for export to the Middle East, European countries, USA, Pakistan and Myanmar.

The above-mentioned cultivars have some salient properties (merit and demerit) such as, yield, taste, disease resistant, tolerant of unfavourable environment, etc., which is important for their cultivation and marketing. The merits and demerits of these cultivars are summarized in the table (Table 6).

From the above properties of merits and demerits, it is apparent that all favourable and beneficial characteristics of vines are not found together in single cultivars or varieties. The betel vine cultivars having good taste, less fibrous, high essential oil content, high yield, disease-resistant, tolerant of unfavourable environment, preservation efficiencies of leaf are not found in nature. For developing high beneficial betel vine farming practice, the mixed or hybrid and improved varieties of the betel vine can be developed in near future considering and incorporating all favourable and advantegeous characteristics of betel vine through biotechnological approaches. In this respect, many scientists are engaged in developing such kinds of betel vine cultivars using the biotechnological approaches for extensive and beneficial farming. Some scientists carried out crossing experiments between male and female vines of Kappori varieties in a Betel vine Research Institute, Bangalore, India. As a result of this experiment, about 60% plants are surprisingly able to produce fruits and seeds, which are also able to grow new betel vine plants. Moreover, it can also be mentioned herein that more intensive and advance researches are required for proper characterization, categorization, quality improvement and conservation of valuable betel vine plants.

REFERENCES

Balasubrahmanyam, V. R., Johri, J. K., Rawat, A. K. S., Tripathi, R. D. and Chaurasia, R. S. (1995) *Betelvine (Piper betle L.).* National Botanical Research Institute Publication, Lucknow, India.

Gamble, J. B. (1928). *Flora of the Presidency of Madras. Botanical Survey of India,* Howrah, India, Vol. II. (Rep. ed. 1957).

Guha, P. (1997). Paan Theke Kutir Silpa Sambhabana" (In Bengali). In: Krishi, Khadya-O- Gramin Bikash Mela, Agricultural and Food Engineering Department, IIT, Kharagpur, India, pp. 15-19. [Guha, P. (1997). Exploring Betel Leaves for Cottage Industry. In: Festival on Agriculture, food, and development of village, Agricultural and Food Engineering Department, IIT, Kharagpur, India, pp. 15-19.]

Guha, P. (2006). Betel leaf: the neglected green gold of India. *Journal of Human Ecology,* 19(2): 87-93.

Khadke, G. N., Hima Bindu, K. and Ravishankar, K. V. (2012). Development of SCAR marker for sex determination in dioecious betelvine (*Piper betle* L.). *Curr Sci*, 103(6): 712-716.

Lakshmi, S. B., and Naidu, K. C. (2010). Comparative Morphoanatomy of *Piper betle* L. cultivars in India. *Annals of Biological Research*, 1(2): 128-134.

Maity, S. (1989). Extension Bulletin: The Betelvine. All India Coordinated Research Project on Betelvine, Indian Institute of Horticultural Research, Hessarghatta, Bangalore, India.

Patra, A. P., Mukherjee, A. K. and Acharya, L. (2011). Comparative study of RAPD and ISSR markers to assess the genetic diversity of Betel vine (*Piper betle* L.) in orissa, India. *Am. J of Biochem Mol Biol,* 1(2): 200-211.

Ranade, S. A., Soni, A., and Kumar, N. (2011). SPAR profiles for the assessment of genetic diversity between male and female landraces of the dioecious betelvine plant (*Piper betle* L.). In: Oscar Grillo (Ed.), *Ecosystems Biodiversity,* pp. 443-464.

Ravindran, P. N., Nirmal Babu, K., Sasikumar, B. and Krishnamoorthy, K. S. (2000). *Botany and crop improvement of black pepper.* In: P N Ravindran (ed.). Black pepper, Piper nigrum. Harwood Academic Publishers, Amsterdam, The Netherlands, pp. 23-142.

Rawat, A. K. S. and Balasubrahmanyam, V. R. (1988). Betelvine cultivars need to be classified. *Ind. Hort.*, 32: 4–6.

Samanta, C. (1994). Paan chaser samasyabali-o-samadhan: Ekti samikkha (In Bengali) [A Report on the Problems and Solutions of Betel Vine Cultivation]. In: Adhikari, H. R., Salt Lake City, Kolkata, India.

Verma, A., Kumar, N. and Ranade, S. A. (2004). Genetic diversity amongst landraces of a dioecious vegetatively propagated plant, betelvine (*Piper betle* L.). *Journal of Biosciences,* 29(3): 319-328.

Chapter 4

PROPAGATION AND GROWTH

1. INTRODUCTION

Propagation of the betel vine is vegetative in nature. It can easily grow and propagate on the soil surface without any support plants or materials in a favourable environmental condition. A rapid growth is commonly found when it gets support plants or materials. The important environmental conditions such as, soil properties, humidity and temperature, irrigation and fertilization, incident of solar radiation, air volume, etc. significantly play pivotal roles in growth and propagation of the betel vine. All of these influencing factors are also directly and indirectly responsible for leaf quality, which in turn influences on the cost-benefit aspects of large-scale and commercial betel vine farming.

2. PROPAGATION

The betel vine is grown by vegetative propagation from the cuttings of stem under partially shaded and humid environment (Figure 17). The cutting or setts can be obtained from past year grown vines having a minimum length of about 1.0 to 2.0 m. The cutting of terminal or apical portions of the vines are easy to grow roots and hence best for planting. Each cutting should contain 2 or more nodes for well and rapid growth. The cutting of one node can also be easily grown under suitable environment. Furthermore, the cuttings from 3 to 5 years-old vines are appropriate for using in propagation in order to get absolute and better results in cultivation of commercial farming.

Generally, bed for planting the cutting of the betel vine is properly prepared by applying fertilizers and treating with anti-pathogenic agents. Setts with vigorous apical buds and nodal adventitious roots are selected and planted at the base of the support materials or live supports, which are planted 4 to 5 months earlier. The cutting is planted in the furrows (8 - 10 cm depth) of sterilized soil with spacing of 50 to 60 cm × 10 to 20 cm during rainy or autumn season. In intensive cultivation, a narrow walking space for work is left between two beds (rows) and bed is slightly higher than the walking space. The width of the walking space between two rows is generally varied from 20 to 45 cm as shown in Table 7 and Figure 18. It can be varied according to types of cultivars and plantation programme of farmers.

The number of cutting required for plantation in a specific area is varied according to the agro-climatic environment and cultivation strategies followed by the farmers in different parts of the world. This number is largely depended on the distance between two rows and between two planted cuttings. On an average 100000 setts are required for planting one hectare area. The number of cutting can be varied with the space left between two beds/rows which are shown in Table 7.

Table 7. Number of cutting generally required per hectare area for planting

Row spacing	Vines/hectare	
	Single vine	Double vine
20 cm	50000	100000
30 cm (1 ft)	30000	60000
45 cm (11/2 ft)	22500	45000

Figure 17. The vegetative propagation of the betel vine.

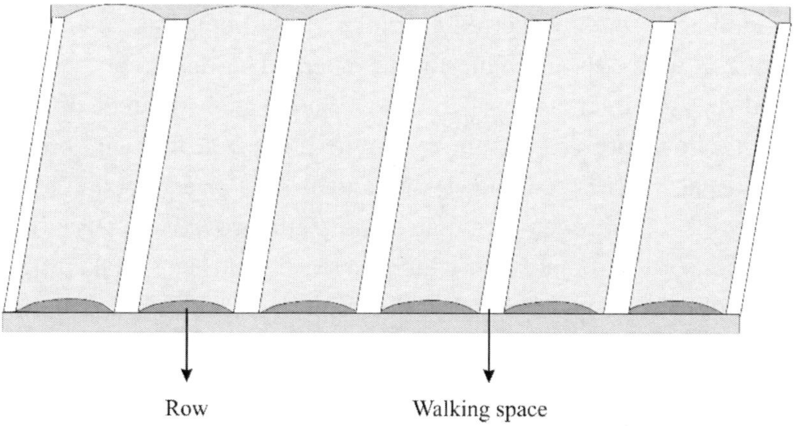

Figure 18. The walking space between the rows.

Figure 19. The vigorous growth of betel vines on live plants (a) and on support materials (b).

3. Growth

The betel vines vigorously grow on living plants and support materials with many branches, and especially male vines grow fast up to height of 20 m (Figure 19). The growth of the betel vine is significantly depended on soil quality, cultivation strategies, agro-climatic conditions, seasons and supporting living plants or materials required for climbing. Growth of the vines is fast during rainy season and slow during winter. It may climb the support trees as high as 10 to 15 ft in a year, although it often grows as an understory ground cover. The betel vines grow to a height 150 to 180 cm in about 3 to 6 months period in closed environmental condition. The branching is noticed in the vines at this stage. The results of 2-year study revealed that the integration of inorganic and organic sources of nutrients (nitrogen, phosphorus, potassium) gave the highest growth and production (Thomas et al., 2013; Chandini, 1989), and also reported that application of higher levels of nitrogen stimulates the high growth because of rapid proliferation of the root system which increases nutrients intake efficiency in the betel vine. Phosphorus and Potassium is absorbed and utilized in greater amounts at higher concentration of nitrogen (Chandini, 1989; Thomas et al., 2013).

4. Factor Influencing Growth

4.1. Soil

It is a most important factor for growth of plant. Allthough, the betel vines grow well in loam or clay loam soil, it has the ability to grow in a wide range of soils from sandy loam to clayey loam (Agri Farming, 2015). The lateritic and clay loam soils are highly suitable for betel cultivation. The soil of friable, organic enriched and has the good drainage capacity with a pH of about 5.6 to 8.2 is favourable for good growth (CSIR, 1969;

Guha and Jain, 1997; Guha, 2006). The soil with low water permeability is not favourable for the problem of water logging.

4.2. Irrigation and Fertilization

Irrigation and fertilization importantly influence the growth of the betel vine. For maintaining the humidity and temperature, light and frequent irrigations are favourable for proper growth of the betel vine, since it is very sensitive to stagnant water. It depends on the season and weather conditions, such as irrigation is not required in rainy season, where as a frequent irrigation is required in winter season. Irrigation is done in such a way that water falls on the leaves and then trickles down to the soil (Kumar, 1999). It also helps to reduce the temperature of air and leaf.

Fertilization is one of the important factors for quantitative and qualitative growth of the betel vine. Generally, organic fertilizer, such as - oil cake, farm yard manure (FYM), etc. can meet all nutrients (such as, nitrogen, phosphorus, potassium, etc.) requirements of the betel vine for good growth. Sometimes, inorganic fertilizer can be applied based on soil quality and necessity of plants.

4.3. Humidity and Temperature

Humidity and temperature are two critical factors for regulating the growth and propagation of the betel vine (Kumar, 1999). High humidity (about 70 - 80%) and mild temperature (about 25 - 30°C) are favourable for high growth. Retarded growth or no growth of the betel vine is observed at a low temperature generally in winter season. Therefore, the design of the betel vine yard should be such to maintain optimum condition of these two important factors (Kumar, 1999).

4.4. Incident of Solar Radiation

Incident of solar radiation on the vine is another important parameter for the growth and propagation of the betel vine. The heating effect of solar radiation can be effectively reduced by blocking the amount of light falling on a surface (Kumar, 1999). Since, it also directly influences the temperature of the cultivated environment. The top shade is required to maintain the incident of solar radiation. The shade of the top can be such, which is able to block 70 - 80% and 40 - 50% incident of solar radiation in summer season and rainy season, respectively (Kumar, 1999).

4.5. Air Volume

Air volume within the system is also significant growth influencing factors in the betel vine cultivation, since it is associated with the regulation of humidity and temperature of the system. If the air volume is large due to big structure of cultivation, more water may be required for maintaining the optimum humidity and temperature of system. That is why; the height of the system should be within 2 – 4 meters to maintain the air volume (Kumar, 1999). The air volume can be changed by increasing and decreasing the height of the system in different seasons as necessary for maintaining effective humidity and temperature.

REFERENCES

Agri Farming (2015). http://agrifarming.in/betel-leaf-farming-information.

Chandini, S. (1989). Management practices for betelvine (*Piper betel* L.). Ph.D thesis, Kerala Agricultural University, Thrissur. India, p p. 130.

CSIR (Council of Scientific and Industrial Research) (1969). The Wealth of India. 8: 84-94. CSIR, New Delhi, India.

Guha P. 2006. Betel leaf: The neglected green gold of India. *Journal of Human Ecology,* 19(2): 87-93.

Guha, P. and Jain, R. K. (1997). Status report on production, processing and marketing of betel leaf (*Piper betle* L.). Agricultural and Food Engineering Department, IIT, Kharagpur, India.

Kumar, N. (1999). Betelvine (*Piper betle* L.) cultivation: A unique case of plant establishment under anthropogenically regulated miclimatic conditions. *Indian journal of History,* 34(1): 231-239.

Thomas, U. C., Chandini, S., and Thomas, A. (2013). Integrated nutrient management studies in betelvine *(Piper betel L.). International Journal of Applied Research and Studies,* 2(7): 1-9.

Chapter 5

CHEMICAL PROPERTIES

1. INTRODUCTION

Betel vine is one of the important medicinal plants due to having several medicinal and therapeutic properties. It has been highly investigated for medicinal and herbal applications by exploring active phytochemicals or biomolecules. The major chemical constituents of betel vine leaf are - water (85 - 90%), proteins (3 - 3.5%), fat (0.4 - 1%), carbohydrates (0.5 - 6.1%), fibre (2.3%), minerals (2.3 - 3.3%), essential oil (0.08 - 0.2%), tannin (0.1 - 1.3%) and alkaloid (arakene) (Pradhan et al., 2013). It also contains all most all vitamins, such as vitamin-C (0.005 - 0.01%), nicotinic acid (0.63 - 0.89 mg/100 gm), vitamin-A (1.9-2.9 mg/100 gm), thiamine (10 – 70 µg/100 gm), riboflavin (1.9 - 30 µg/100 gm). The protein part of betel leaf is formed by almost all kinds of amino acids as well as enzymes like diastase and catalase, which are found only in traces (CSIR, 1969; Gopalan et al., 1984; Guha and Jain, 1997). Betel leaf contains 0.38% to 1.46% reducing sugars. Minerals, such as calcium (0.2 - 0.5%), iron (0.005 - 0.007), Idoine (3.4 µg/100gms), phosphorus (0.05 - 0.6%), potassium (1.1 - 4.6%), etc. are also the chemical constituents of the leaf (Guha, 2006). Some percentage (about 0.7 - 2.6%) of bitter compounds are present in leaf.

The aromatic value is determined by the nature of oil content in leaf. Oil content varies from 0.7 to 2.6%. The characteristically strong pungent aromatic flavour of leaves is mainly due to phenol and terpene like compounds (Bajpai et al., 2010). The entire phenol content is found to be varied with the gender of the betel vine. The male plant contains three-fold higher total phenols content and two-fold higher thiocyanate content as compare to female plant. The quality of the leaf depends upon the phenolic content, i.e., the more phenolic content the better leaf quality (Bissa et al., 2007).

2. Biochemical Properties

Recently, a number of scientists reported that betel leaves contain starch, diastases, sugars and an essential oil composing of safrole, allyl pyrocatechol monoacetate, eugenol, terpinen-4-ol, eugenyl acetate, etc. as the major components (Chopra and Chopra, 1958; Kanjwani et al., 2008). Phytochemical analysis of leaves revealed the presence of alkaloids, carbohydrate, amino acids, tannins and steroidal components (Sugumaran et al., 2011). The middle part of the main vine contains largest quantity of tannin. The terpenoids include 1, 8- cineole, cadinene, camphene, caryophyllene, limonene, pinene, cavicol, ally pyrocatechol, carvacrol, safrole, eugenol and chavibetol, are the major phenols found in betel leaf. Eugenol was identified as the antifungal compound in the oil. The fresh new leaves contain much more amounts of essential oil diastase enzyme and sugar as compare to old leaves. Chavicol is four times potent as an antiseptic agent as compare to carbolic acid.

It contains a wide array of biologically active compounds whose concentration depends upon the variety of betel plant, season and climate (Rai et al., 2011). Several phytochemicals found in the betel vine are: chavibetol, chavicol, hydroxychavicol, estragole, eugenol, methyl eugenol, hydroxycatechol, caryophyllene, eugenol methyl ether, cadinene, γ-lactone, allyl catechol, p-cymene, cepharadione A, dotriacontanoic acid, tritriacontane, p-cymene, terpinene, eucalyptol, carvacrol, sesquiterpenes, cadinene, caryophyllene, dotriacontanoic acid, hentriacontane, pentatriacontane, stearic acid, n-triacontanol, triotnacontane, piperlonguminine, allylpyrocatechol diacetate, isoeugenol, 1, 8-cineol, α-pinene, β-pinene, sitosterol, β-sitosteryl palmitate, γ-sitosterol, stigmasterol, ursolic acid and ursolic acid 3β-acetate (Rastogi and Mehrotra, 1993; Kumar et al., 2010; Rai et al., 2011). Some phyto-chemicals such as – ugenol, α-tocopherol, ursolic acid, Hydroxychavicol and β-carotene are most important anticarcinogenic compounds found in the betel vine (Rai et al., 2011).

Despite these above compounds, the plant contains inorganic elements like Mg, Si, Cl, K, Ca, Zn, which is determined by the eergy dispersive X-ray analysis on the leaf of *P. betle* (L) var. SGM1 (Mubeen et al., 2014).

The name and structure of some phyto-constituents identified in the betel vine are shown as below (Kumar et al., 2010; Chaurasia et al., 2010; Ghosh and Bhattacharya, 2005; Fong, 2009; Chandra et al., 2011; Arambewela et al., 2005, 2011; Dwivedi and Meheta, 2011; Dwivedi et al., 2010; Pradhan,et al., 2013):

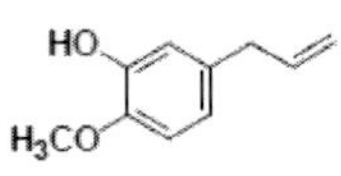

Chavibetol

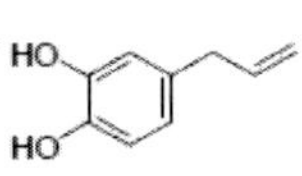

Allypyrocatechol

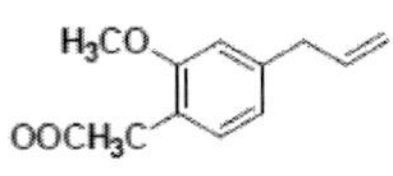

Chavibetol acetate

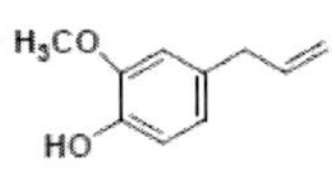

Eugenol

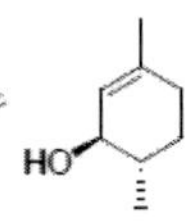

Piperitol

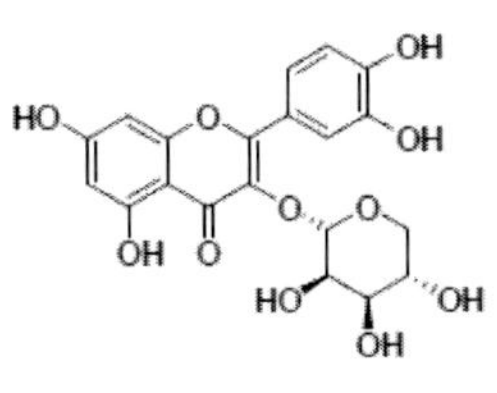

Quercetin

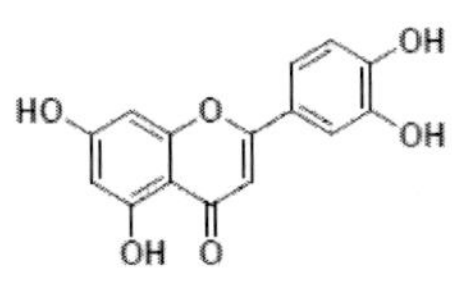

Luteolin

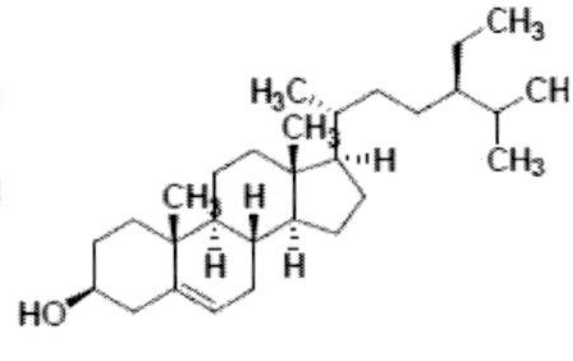

β- sitosterol

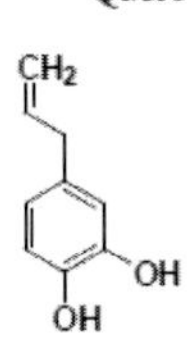

Hydroxychavicol

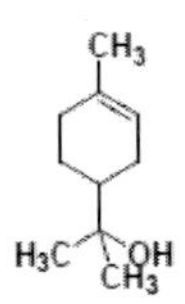

α-terpineol

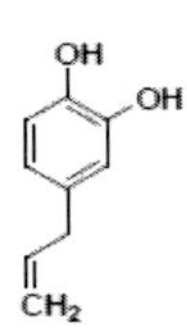

Allyl catecol

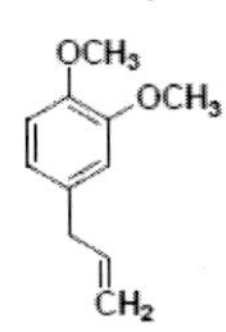

Eugenol methyl ether

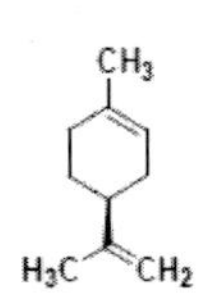

D- limonene

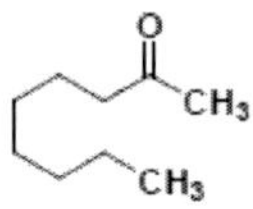
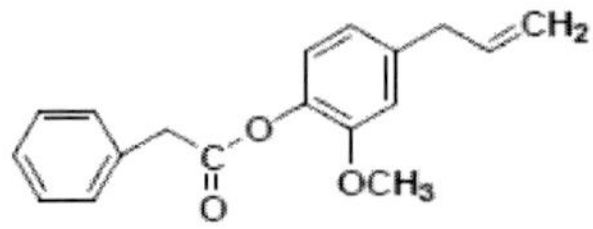
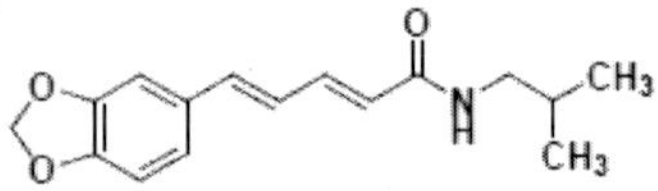

2-noanone 4-allyl phenyl acetate (db) Piperlonguminine

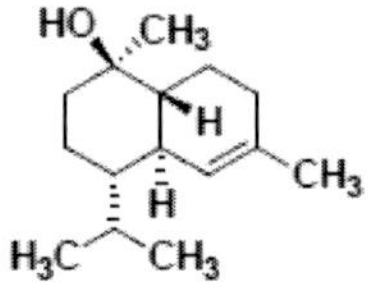
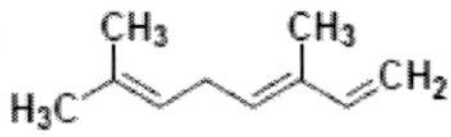
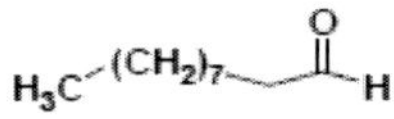
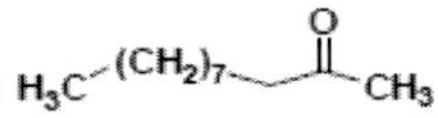

α-cadinol Ocimene N-decanal 2-undecanone

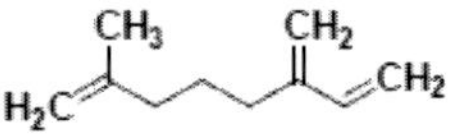
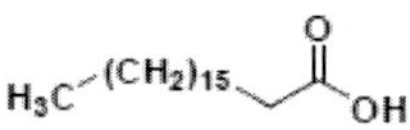
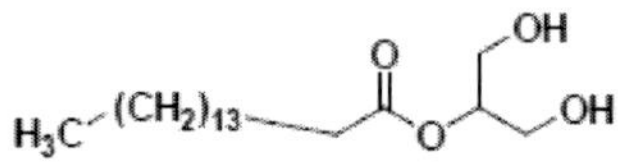

Myrcene Stearic acid 2- Mono palmitin

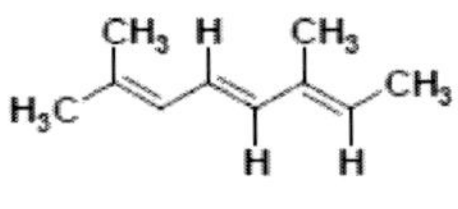
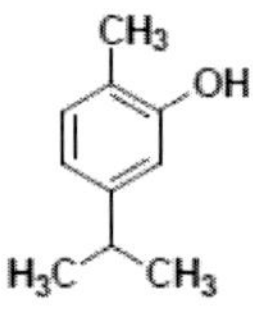
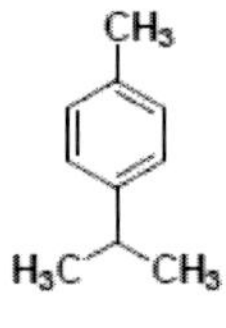
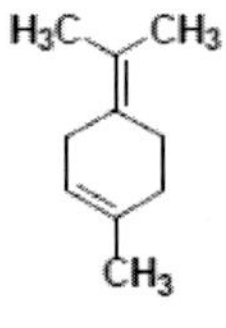
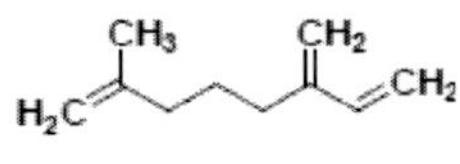

Allo ocimene Cavacrol Cymene Terpinolene α-Myrcene

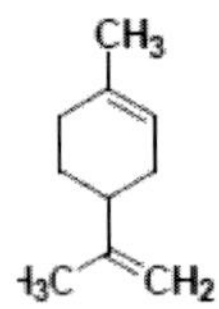
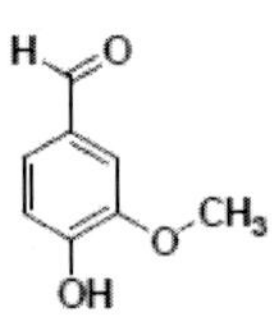
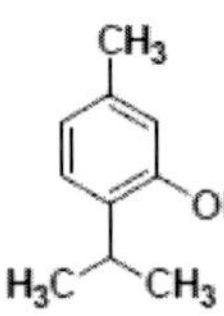
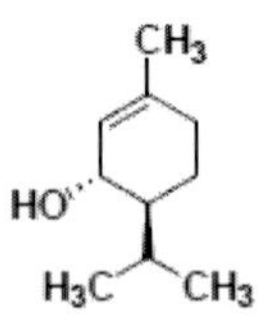
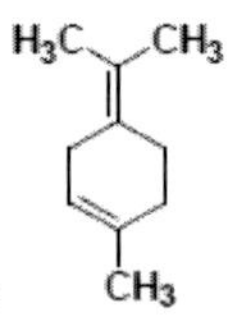
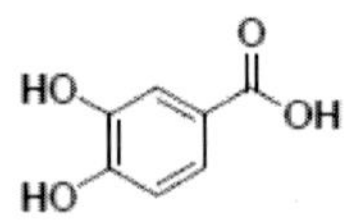

Limonine Vanillin Thymol Cis-piperitol Terpinolene Procatechuic acid

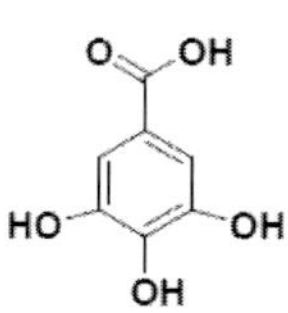
Gallic acid

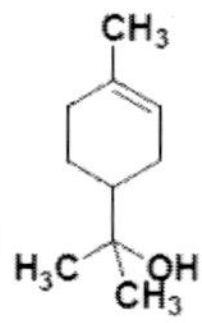
α-terpineol

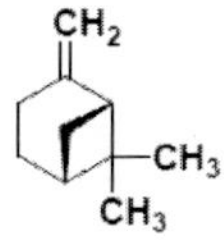
β-pinene

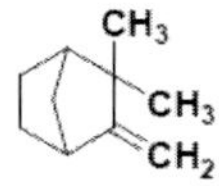
Camphene

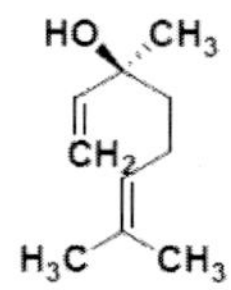
Linalool

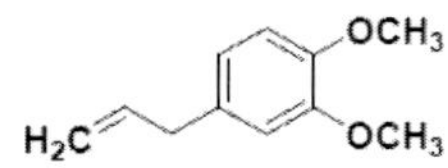
Allyl diacetoxy benzene

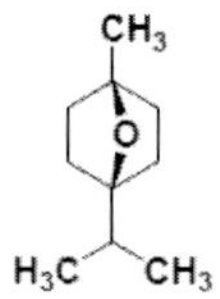
Eucalyptol

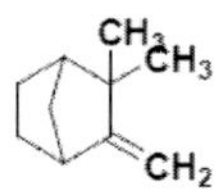
Camphene

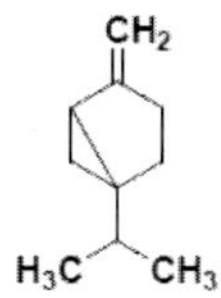
Sabinene

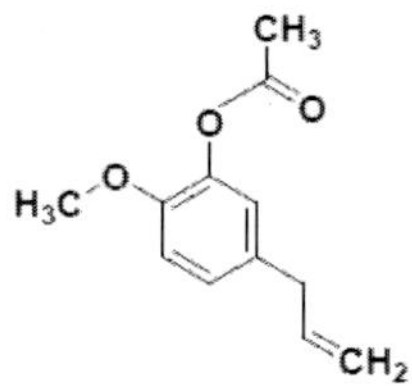
3-allyl-6-methoxyphenolb

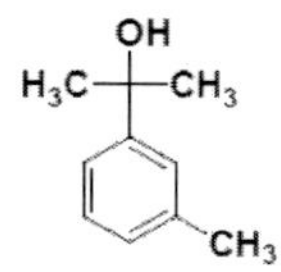
m-Cymen-8-ol

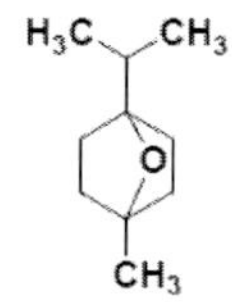
1, 4 cineole

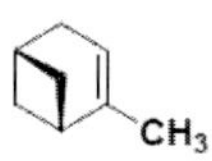
α-pinene

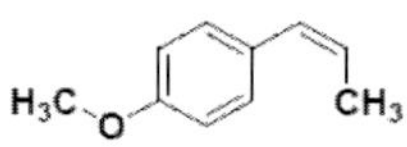
Anethole

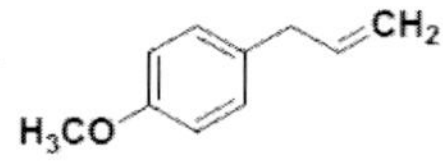
Estragol

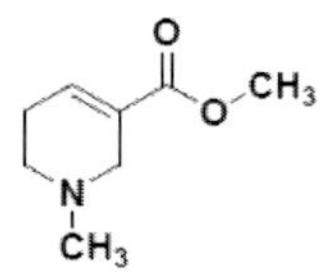
Arecoline

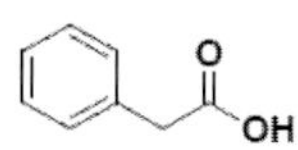
Benzene acetic acid

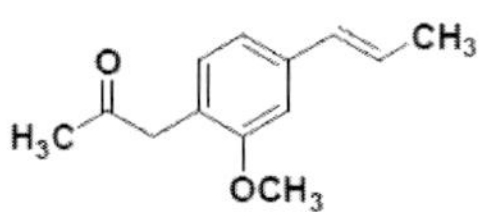
Iso eugenyl acetate

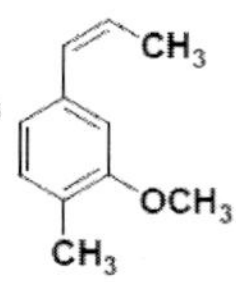
Isoeugenol

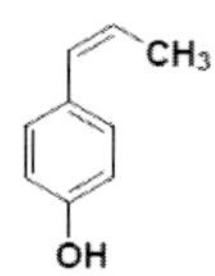
Chavicol

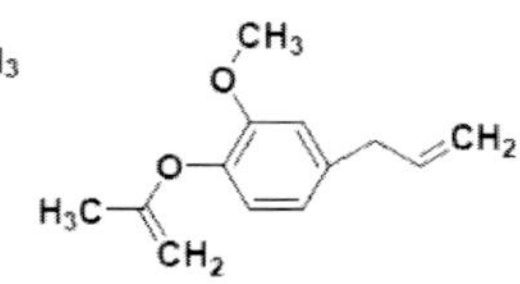
Eugenyl acetate

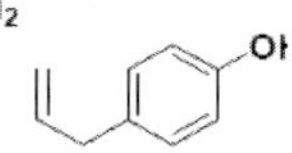
4-allyl phenol

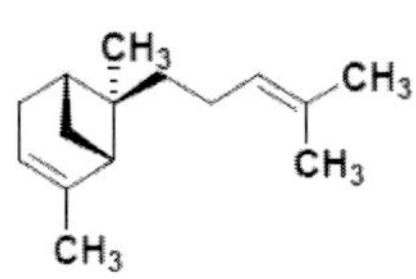
α-bergamotene

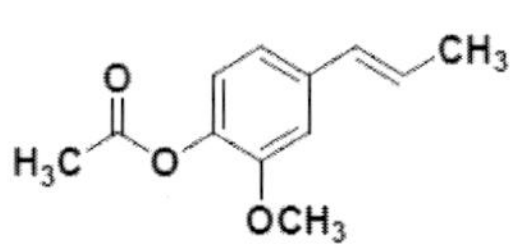
Isoeugenyl acetate

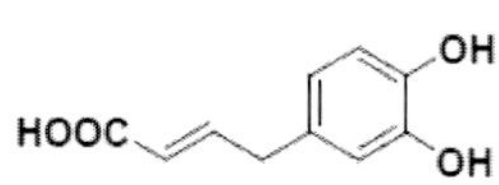
Caffeic acid

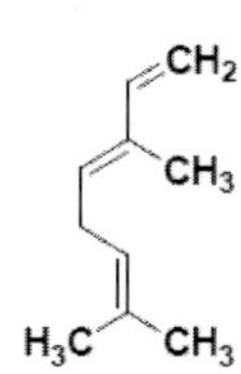
(E)-ß-ocimene

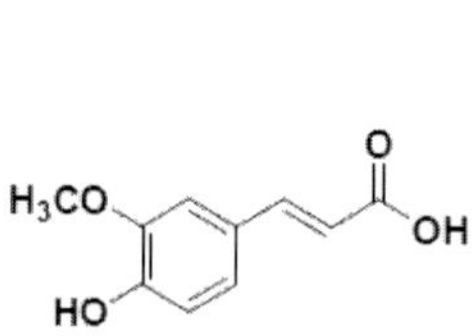

Ferulic acid

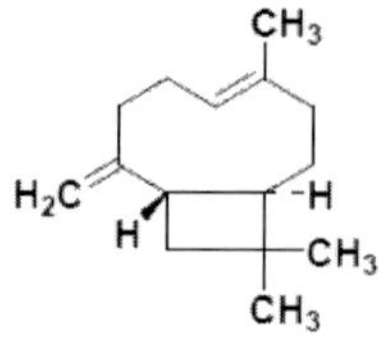

Carryophyllene

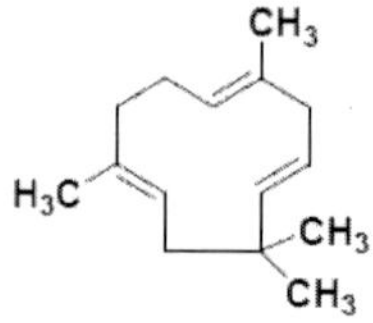

Humlene

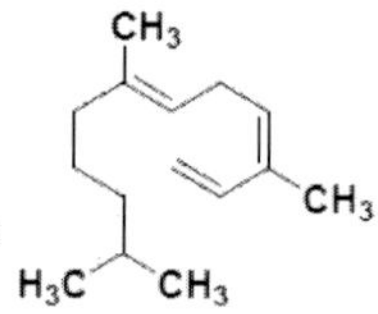

α-farnesene

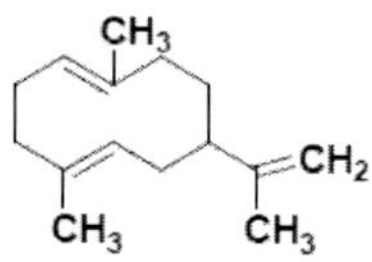

Germacrene-A

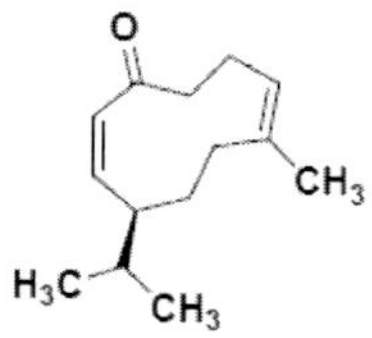

Germacrene-D

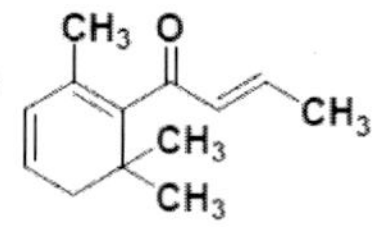

(E)-ß- Damascenone

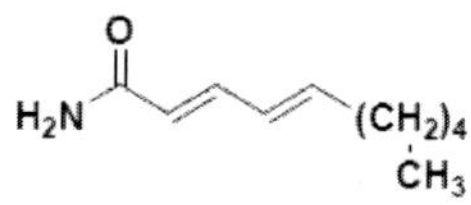

4E-decadienamide

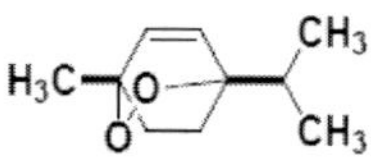

Isoascaridole

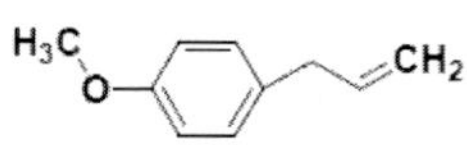

4-Allyl anisole

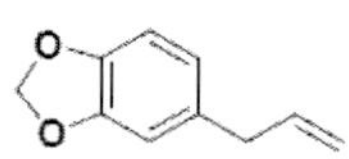

Safrole

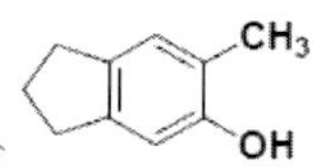

5-Indanol

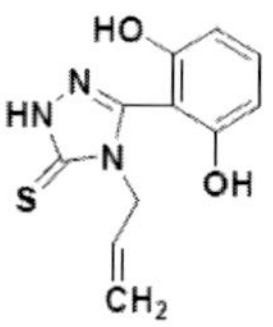

4-allyl resorcinol

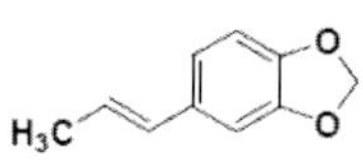

ß-iso safrole

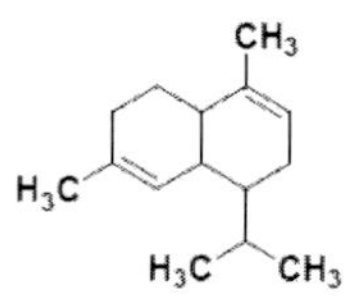

α-muurolene

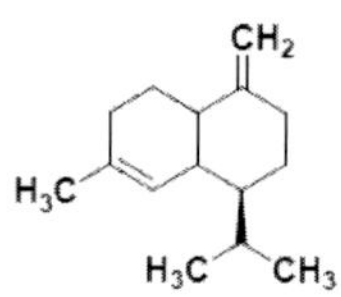

Cadinene

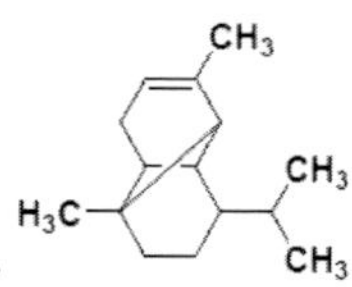

α- copaene

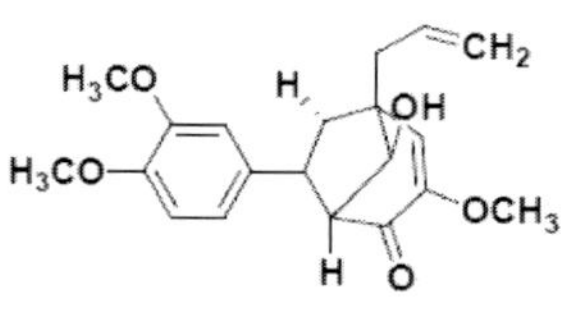

Piperol-A

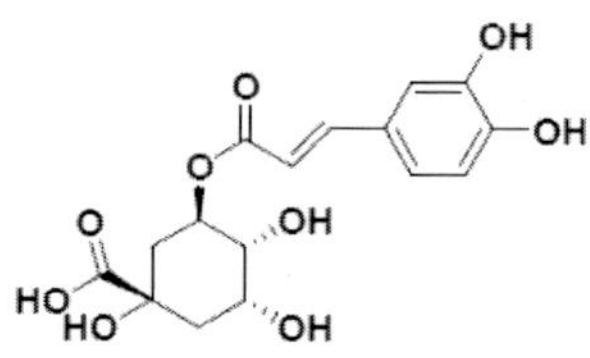

Chlorogenic acid

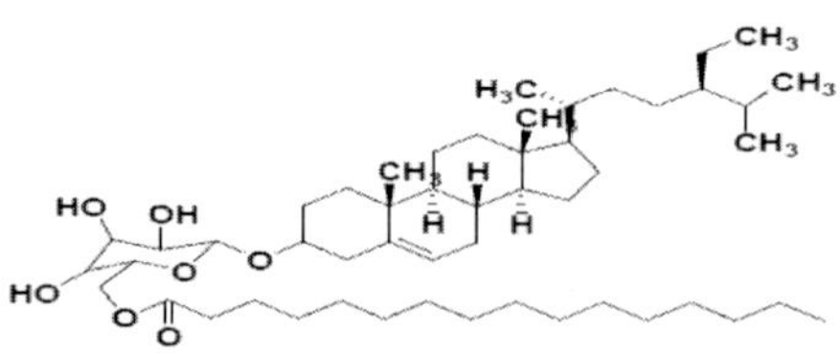

ß-sitosteryl palmitate

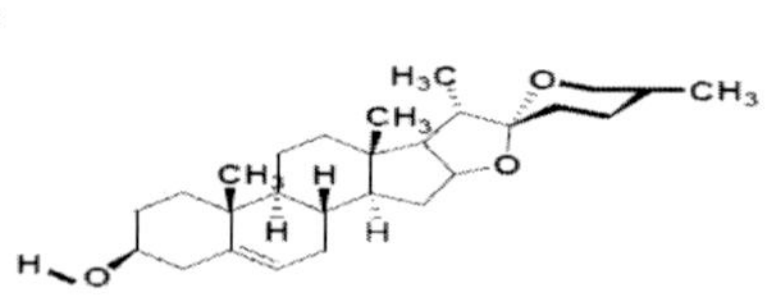

Diosgenin

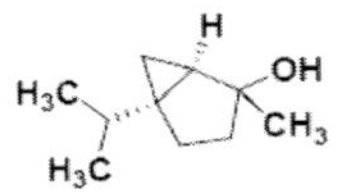
Cis Sabinene hydrate

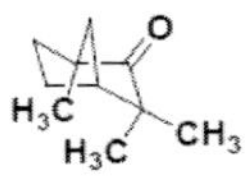
Fenchone

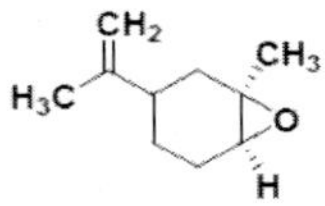
Cis-Limonene oxide

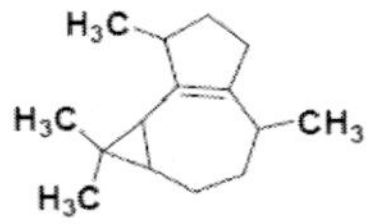
Isoledene

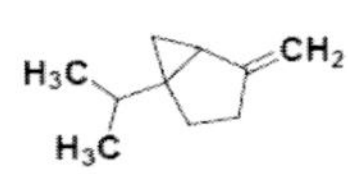
Sabinene

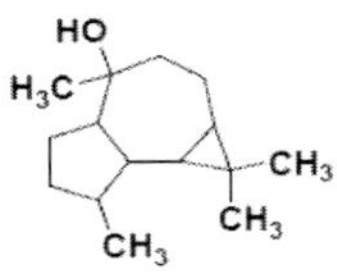
Ledol

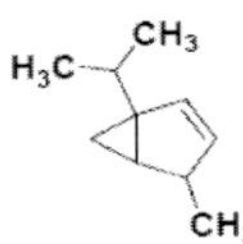
Thujene

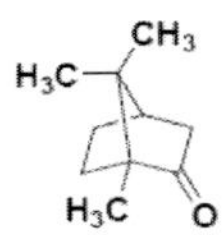
Camphor

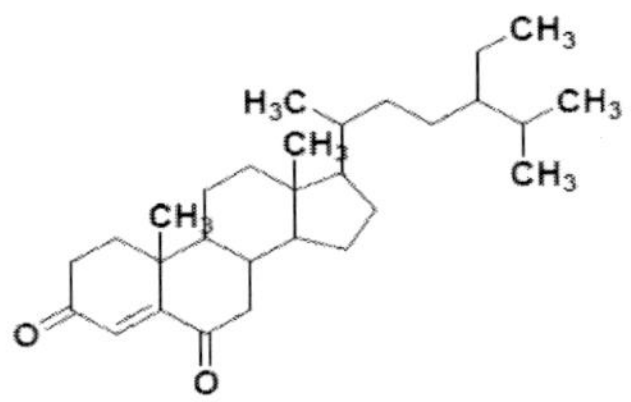
Stigmast-4-en-3,6-dione

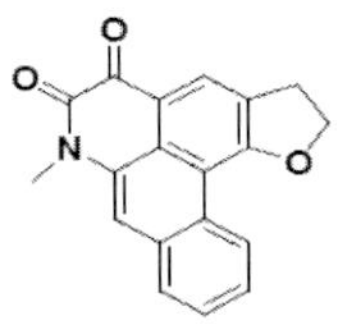
Cepharadione A

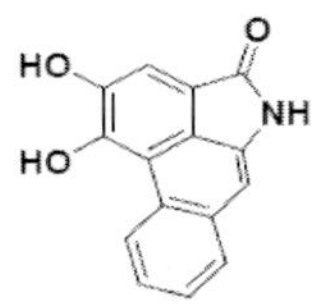
Aristololactame (A-II)

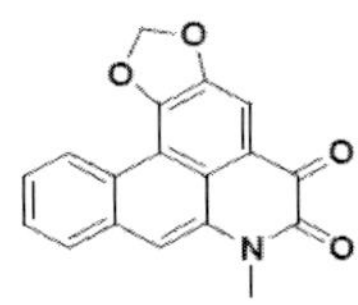
Cepharadione

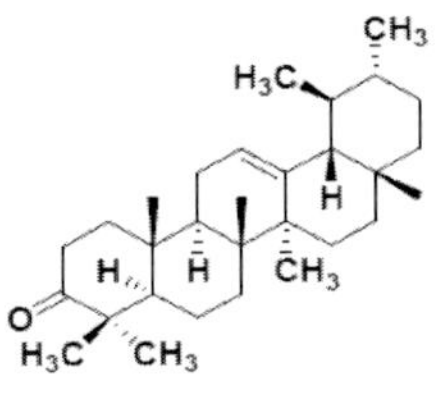
Ursonic acid

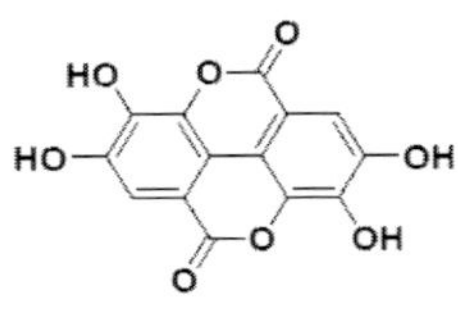
Ellagic acid

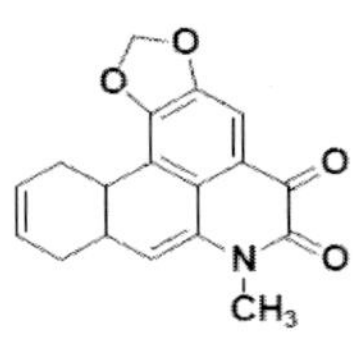
Cepharadione-A

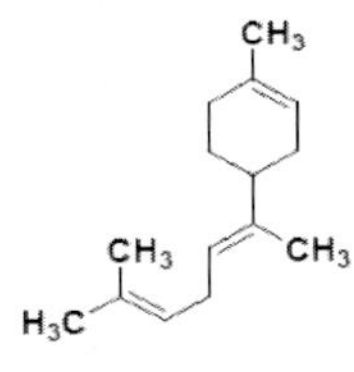
α-Bisabolene

However, relevant data from a complete biochemical analysis is not available from any single source so far. Because, the analytical values from two or more reports may reveal a very wide or even contradictory trend, therefore, more research work in this direction becomes inevitable.

Besides, many advanced and analytical results obtained from current researches are also able to reveal the details of biological activities and action mechanisms of some bio-molecules identified in *P. betle* (Table 8) (Kumar et al., 2010).

Table 8. Specific biological activities and action mechanism of some bio-molecules of *Piper betle*

Chemical constituent	Specific biological activity	Mode of action	Reference
Hydroxychavicol (HC)/Hydroxychavicol acetate (HCA)	Hyperuricemia (antidiabetic)	Acts via xanthine oxidase inhibition	Murata et al., 2009
	Cytokine production in Th cells (increased IL-2 production and attenuates IFN-gamma expression in Th cells) (immunomodulatory)	Suppressed T-bet expression, which is responsible for IL-2 suppression and IFN-gamma induction in Th cells and inhibited T-bet-mediated Th1 cell differentiation	Min, 2009
	Oral hygiene	Probably works through the disruption of the permeability barrier of microbial membrane structures	Sharma et al., 2009; Ali et al., 2010
	Inhibits platelet aggregation	A potent COX-1/COX-2 inhibitor, ROS scavenger and inhibits platelet calcium signalling, TXB(2) production and aggregation	Chang et al., 2007
	Chemopreventive against the tobacco-specific carcinogens	Suppressed the mutagenic effects of tobacco-specific *N*′-nitrosonornicotine and 4-(nitrosomethylamino)-1-(3-pyridyl)-1-butanone	Amonkar et al., 1986; Amonkar et al., 1989
	Oral KB carcinoma cells	Inhibits the growth, adhesion and cell cycle progression of KB cells, whereas induction of KB cell apoptosis (HC > 0.1 mM) was accompanied by cellular redox changes	Chang et al., 2002
Allylpyrocatechol (APC)	Gastric ulcer-healing action	Mediated by modulation of arginase metabolism and shift of cytokine balance	Yadav et al., 2009
	Protection against ulceration	Protects indomethacin-induced gastric ulceration due to its antioxidative and mucin-protecting properties	Tripathi, 2008

Chemical constituent	Specific biological activity	Mode of action	Reference
	Anti-inflammatory effect	Targets the inflammatory response of macrophages via inhibition of iNOS, COX-2 and IL-12 p40 through down-regulation of the NF-kappa B pathway, indicating that APC may have therapeutic potential in inflammation associated disorders	Santhakumari et al., 2006
Chavibetol (CHV)	Photoprotective /radioprotective	Protects photosensitization-mediated lipid peroxidation of rat liver mitochondria; prevents γ -ray induced lipid peroxidation as assessed by measuring TBARS	Rathee et al., 2006; Daniell et al., 2009; Shojaei, 1998
Piperbetol Methylpiperbetol Piperol A Piperol B	Platelet hyperactivity/cardiovascular diseases due to intravascular thrombosis	Selectively inhibited platelet aggregation induced by platelet activating factor (PAF) in a concentration dependent manner. These constituents are effective PAF receptor antagonists *in vitro*. These phenols had no effects on the cAMP contents in resting rabbit platelets.	Zeng et al., 1997

REFERENCES

Ali, I. et al., (2010). *In vitro* antifungal activity of hydroxychavicol isolated from *Piper betle* L. *Ann. Clin. Microbiol. Antimicrob.*, 9: 7.

Amonkar, A. J., Nagbhushan, M., D'Souza, A. V. and Bhide, S. V. (1986). Hydroxychavicol: a new phenolic antimutagen from betel leaf. *Food Chem. Toxicol.*, 24: 1321–1324.

Amonkar, A. J., Padma, P. R. and Bhide, S. V. (1989). Protective effect of hydroxychavicol, a phenolic component of betel leaf, against the tobacco-specific carcinogens. *Mutat. Res.*, 1989, 210: 249–253.

Arambewela, L. S. R., Arawwawala, L. D. A. M., Kumaratunga, K. G., Dissanayake, D. S., Ratnasooriya, W. D. and Kumarasingha, S. P. (2011). Investigations on *Piper betle* grown in Sri Lanka. *Pharmacogn Rev.*, 5(10): 159–163.

Arambewela, L., Kumaratunga, K. G. A. and Dias, K. (2005). Studies on *Piper betle* of Sri Lanka. *J.Natn.sci.Foundation Sri Lanka,* 33(2): 133-139.

Bajpai, V., Sharma, D., Kumar, B. and Madhusudanan, K. P. (2010). Profiling of *Piper betle* Linn. cultivars by direct analysis in real time mass spectrometric technique. *Biomedical Chromatography*, 24(12): 1283-1286.

Bissa, S., Bohra, A. and Songara, D. (2007). Traditions in oral hygiene: Chewing of betel (*Piper betle* L.) Leaves. *Scientific correspondence*, 92(1): 24-28.

Chandra, V., Tripathi, S., Verma, N. K., Singh, D. P., Chaudhary, S. K. and Roshan, A. (2011). *Piper betel*: phytochemistry, traditional use & pharmacological activity-a review. *IJPRD*, 4(4): 216 -223.

Chang, M. C. et al., (2007). Hydroxychavicol, a novel betel leaf component, inhibits platelet aggregation by suppression of cyclooxygenase, thromboxane production and calcium mobilization. *Br. J. Pharmacol.*, 152: 73–82.

Chang, M. C., Uang, B. J., Wu, H. L., Lee, J. J., Hahn, L. J. and Jeng, J. H. (2002). Inducing the cell cycle arrest and apoptosis of oral KB carcinoma cells by hydroxychavicol: roles of glutathione and reactive oxygen species. *Br. J. Pharmacol.*, 135: 619–630.

Chaurasia, S., Kulkurni, G. T. and Setty, L. N. (2010). Phytochemical studies and invitro cytotoxicity. *International Research Journal of Pharmacy*, 1(1): 384-391.

Chopra, R. N. and Chopra, I. C. (1958). Indigenous Drugs of India. Pub- Academic Publishers 2nd Edition. pp. 372.

CSIR (Council of Scientific and Industrial Research, New Delhi) (1969). *The Wealth of India,* CSIR, New Delhi, 8: 84-94.

Daniell, H., Singh, N. D., Mason, H. and Streatfield, S. J. (2009). Plantmade vaccine antigens and biopharmaceuticals. *Trends Plant Sci.*, 14: 669–679.

Dwivedi, B. K. and Mehta, B. K. (2011). Chemical investigation of aliphatic compounds of *Piper betle* (leaf stalk). *J. Nat. Prod. Plant Resour.*, 1(2): 18-24.

Dwivedi, B. K., Kumar, S., Nayak, C. and Mehta, B. K. (2010). Gas chromatography mass spectrometry (GCMS) analysis of the hexane and benzene extracts of the Piper betle (leaf stalk) (Family: Piperaceae) from India. *Journal of MedicinalPlants Research,* 4(21):.2252-2255.

Fong, Y. S., (2009). Chemical constituents and biological activities of Garcinia mangostana and Piper betle. Master's thesis, University Putra Malaysia, pp. 22.

Ghosh, K. and Bhattacharya, T. K. (2005). Chemical Constituents of *Piper betle* Linn. (Piperaceae) roots. *Molecules,* 10: 798-802.

Gopalan, C., Ramasastri, B. V. and Balasubramanian, S. C. (1984). Nutritive Value of Indian Foods. National Institute of Nutrition (ICMR), Hyderabad, India, pp. 108.

Guha, P. 2006. Betel leaf: the neglected green gold of India. *Journal of Human Ecology,* 19(2): 87-93.

Guha, P. and Jain, R. K. (1997). Status Report On Production, Processing and Marketing of Betel Leaf (*Piper betle* L.). Agricultural and Food Engineering Department, IIT, Kharagpur, India.

Kanjwani, D. G., Marathe, T. P., Chiplunkar, S. V. and Sathaye, S. S. (2008). Evaluation of immunomodulatory activity of methanolic extract of *Piper betel*", *Scandinavian Journal of Immunology*, 67: 589–593.

Kumar, N., Misra, P., Dube, A., Bhattacharya, S., Dikshit, M. and Ranade, S. (2010). *Piper betle* Linn. a maligned Pan-Asiatic plant with an array of pharmacological activities and prospects for drug discovery. *Current science,* 99(7): 922-932.

Min, H. J., Nam, J. W., Yu, E. S., Hong, J. H., Seo, E. K. and Hwang, E. S. (2009). Effect of naturally occurring hydroxychavicol acetate on the cytokine production in T helper cells. *Int. Immunopharmacol.*, 9: 448–454.

Mubeen, M., Periyanayagam, K. and Sathik Basha, S. (2014). Anatomical Investigation on the leaves of *Piper betle* (L) var. Sirugamani 1(SGM1) links an Ethnomedical important Medicinal plant and its Pharmacognostic relevance. *International Journal of Pharm Tech Research,* 6(1): 244-251.

Murata, K. et al., (2009). Hydroxychavicol: a potent xanthine oxidase inhibitor obtained from the leaves of betel, *Piper betle*. *J. Nat. Med.*, 63: 355–359.

Pradhan, D., Suri, K. A., Pradhan, D. K. and Biswasroy, P. (2013). Golden Heart of the Nature: *Piper betle* L. *Journal of Pharmacognosy and Phytochemistry*, 1: 147-167.

Rai, M. P., Ramdas, K., Princy, L. P., Rao, P., Rao, S., Bhat, H. P. and Baliga, M. S. (2011). *Piper Betel* Linn (Betel Vine), the Maligned Southeast Asian Medicinal Plant Possesses Cancer Preventive Effects: Time to Reconsider the Wronged Opinion. *Asian Pacific Journal of Cancer Prevention,* 12: 2149 – 2156.

Rastogi, R. P. and Mehrotra, B. N. (1993). Compendium of Indian Medicinal Plants, Vol. 3; Publications and Information Directorate, *New Delhi*, 502-503.

Rathee, J. S., Patro, B. S., Mula, S., Gamre, S. and Chattopadhyay, S. (2006). Antioxidant activity of *Piper betle* leaf extract and its constituents. *J. Agric. Food Chem.*, 54: 9046–9054.

Santhakumari, P., Prakasam, A. and Pugalendi, K. V. (2006). Antihyperglycemic activity of *Piper betle* leaf on streptozotocin-induced diabetic rats. *J. Med. Food*, 9: 108–112.

Sharma, S. et al., (2009). Evaluation of the antimicrobial, antioxidant, and anti-inflammatory activities of hydroxychavicol for its potential use as an oral care agent. *Antimicrob. Agents Chemother.*, 53: 216–222.

Shojaei, A. H. (1998). Buccal mucosa as a route for systemic drug delivery: a review. *J. Pharm. Pharm. Sci.*, 1: 15–30.

Sugumaran, M., Poornima, M., Venkatraman, S., Lakshmi and M., Srinivasansethuvani. (2011). Chemical composition and antimicrobial activity of sirugamani variety of Piper betle Linn Leaf oil. *Journal of Pharmacy Research,* 4(10): 3424-3426.

Tripathi, S., (2008).Chemical investigation of betel vine (*Piper betle* L.) as antioxidant agent. Ph D thesis, Lucknow University, Lucknow, India.

Yadav, S. K., Adhikary, B., Maity, B., Bandyopadhyay, S. K. and Chattopadhyay, S. (2009). The gastric ulcer-healing action of allylpyrocatechol is mediated by modulation of arginase metabolism and shift of cytokine balance. *Eur. J. Pharmacol.*, 614: 106–113.

Zeng, H. W., Jiang, Y. Y., Cai, D. G., Bian, J., Long, K. and Chen, Z. L. (1997). Piperbetol, methylpiperbetol, piperol A and piperol B: a new series of highly specific PAF receptor antagonists from *Piper betle. Planta Med.*, 63: 296–298.

Chapter 6

Importance of Betel Vine

1. Introduction

The betel vine can be considered as one of the important gifts of nature, because of its several importances in different aspects of human health and society as well as in environment health. The significant importance of the betel vine has been proved in each and every plethora of human life from the dawn of civilization. It is cultivated intensively especially for its leaves, which are consumed by about 15 to 20 million people in India. It has immense social and cultural, economic, nutritional and medicinal importances as well as a folk (Siddha and Ayurvedha) reputations in the rural peoples of some countries like - India, Bangladesh, Nepal, Srilanka, Pakistan, Myanmar, Indonesia, Malaysia and Vietnam. According to Guha (2006), the leaf is widely used in social, cultural and religious events for hospitality and also in medicinal purposes. That's why; it is recognized as "green gold" or "green heart" of the nature. People use this valuable plant in various purposes from long back, which has also been proved from *paleontological* evidence by different researchers as mentioned previously elsewhere.

2. Significant Importance

The detail importances of the betel vine are described in the following eight aspects:

1. Social and cultural use
2. Socio-economic importance
3. Nutritional value
4. Culinary uses
5. Medicinal value
6. Industrial value
7. Aesthetic use
8. Environmental importance

2.1. Social and Cultural Use

The betel vine has immense social and cultural importance. The leaf is offered to god as holy or welfare leaf during worship in some societies of India, Bangladesh, Nepal, Vietnam, etc. The plant is also worshiped as a goddess such as - the “Durga” or the “Kali” in some rural peoples of India as a part of ritual cultures. Raw leaves of the betel vine are traditionally used for chewing along with many other ingredients like slaked lime, sliced areca nut, catechu, coriander, aniseed, clove, cardamom, sweetener, coconut scrapings, ashes of diamond, pearl, gold and silver (Ayurvedic preparations), jelly, pepper mint, flavouring agent, fruit pulp, etc. (CSIR, 1969). Chewing of betel leaf is an ancient custom or habit of over 2000 years (Rooney, 1993) for refreshing the mouth. Above chewing combination in the form of betel quid could be varied with different ingredients from country to country (Figure 20). Several hundred million people today practice the ancient custom of betel-chewing. In fact, the edible betel leaf has achieved an esteemed position in the human society rightly from the dawn of civilization, particularly in the countries like Bangladesh, Burma, China, India, Indonesia, Malaysia, Nepal, Thailand, etc. (Jana, 1996; Khoshoo, 1981; Samanta, 1994; Sharma et al., 1996; Guha, 2006).

The leaves have also been used in religious occasions. It is also auspicious to make offering of betel leaf and areca nut on the occasions such as religious ceremonies, festival (pujas) and wedding ceremonies. In India, it is an age old customary to offer and serve betel leaf to guests as a symbol of cultural respect on various social, aesthetic and religious occasions. The betel vine, areca tree and lime pot are considered as family members with deep respects in the culture of some South and South East Asian countries, especially in India, Bangladesh, Nepal, Vietnam, etc.

(a) (b) (c) (d)

Figure 20. Betel quid prepared using different ingredients (a and b) and different forms of betel quid (c and d).

Table 9. Nutritional composition of fresh betel leaf (Guha, 2006)

S. No.	Constituents	Approximate composition	
1	Water	85 – 90%	
2	Protein	3 – 3.5%	
3	Fat	0.4 – 1.0%	
4	Minerals	2.3 – 3.3%	
5	Fibre	2.3%	
6	Chlorophyll	0.01 – 0,25%	
7	Carbohydrate	0.5 – 6.10%	
8	Nicotinic acid	0.63 – 0.89	mg/100 g
9	Vitamin C	0.005 – 0.01%	
10	Vitamin A	1.9 – 2.9	mg/100 g
11	Thiamine	10 – 70	µg/100 g
12	Riboflavin	1.9 – 30	µg/100 g
13	Tannin	0.1 – 1.3%	
14	Nitrogen	2.0 – 7.0%	
15	Phosphorus	0.05 – 0.6%	
16	Potassium	1.1 – 4.6%	
17	Calcium	0.2 – 0.5%	
18	Iron	0.005 – 0.007%	
19	Iodine	3.4	µg/100 g
20	Essential oil	0.08 – 0.2%	
21	Energy	44	kcl/100 g

2.2. Socio-Economic Importance

The betel vine is grown as an important, traditional and ancient cash crop. Its leaf is one of the promising commercial crops capable of attracting substantial amount of foreign exchange to the country (Bhakta et al., 2016). Because, it has good export potential, and India exports huge betel leaves to the other countries. The leaves are also in great demand in several other countries of the world where it is either not grown at all or the demand exceeds the local production and supply. Consequently, leaves are exported to many countries like - Bahrain, Canada, Great Britain, Hong Kong, Italy, Kuwait, Nepal, Pakistan, Saudi Arab and many European countries, USA. The leaves worth about Rupees 30 to 40 million are only exported to European countries. This clearly indicates the foreign exchange earning potentiality of the crop. If merely transportation and marketing facilities, including the export channels are adequately developed, then the revenue generated by this leafy crop would easily exceed than that of any other major crop in the country even with the present level of traditional agronomical practices. In fact, the revenue generated by the crop may be further magnified by many folds if the agronomic practices are scientifically explored.

Betel farming and it's up and down streams' activities can generate employment opportunities throughout the year. Furthermore, as far as the national employment generation is concerned, about 20 million people derive their livelihood directly or indirectly from production, processing, handling, transportation and marketing of betel leaves in India. It is one of the most important cash crops, which uplift the socio-economic status of certain percentage people (Bhakta et al., 2016). This significance justifies its nomenclature as the "green gold of nature". A well-coordinated effort by the farmers, traders, scientists, administrators and policy makers is required to initiate and boost up the national economy through proper exploitation of potentiality of this green gold.

2.3. Nutritional Value

Besides the aesthetic, social and economic value, the betel leaves have nutritional importance, because it consists of the substantial amount of nutrients, vitamins, minerals and other valuable components as mentioned in Table 9 (Guha, 2006). It is also beneficial in accelerating the process of digestion. It is known that six leaves with a bit of slaked lime is said to be comparable to about 300 ml of cow milk, particularly in respect to vitamin and mineral nutrition (Guha, 2006). The enzymes like diastase and catalase are also present in the leaf. Besides, a significant amount of all the essential amino acids are present except lysine, histidine and arginine, which are found only in traces (CSIR, 1969; Gopalan, 1984; Guha and Jain, 1997).

2.4. Culinary Uses

The betel leaf is used as a popular spice in South East Asian cooking. It is also utilized as raw and cooked forms. A traditional culinary use of leaves are wrapping for spiced minced meat and other morsels of food (Figure 21). It is often employed for decorating the food to bring attractiveness of served food, because the leaves are so glossy with attractive shape. The white flower spikes of the betel plant develop into seeds/fruits that look at a little like a green/brown mulberry when ripe and can be eaten; it is a tasty morsel of sweet jelly-like pulp.

(a) (b)

(c) (d)

Figure 21. Figures showing the different culinary applications of betel leaf: (a and b) used for wrapping the prepared food, (c and d) used as an ingredient for preparing food.

2.5. Medicinal Value

The betel vine is called as "green heart of nature". Because of its heart-shaped leaves possessing numerous medicinal importances are directly and indirectly associated with nourishing and saving the human lives, which can be described in the follows:

2.5.1. Traditional Therapy

The betel leaf has been using as a folk medicine in different treatment purposes of the human society right from the dawn of civilization. Folktales reflect on the various application aspects of the betel leaf. Chewing areca nut and betel leaf is a good remedial measure of bad breath. It has been known to use as an aromatic, stimulo-carminative (CSIR, 1992), astringent and aphrodisiac (kamagnisandipanam) agent from ancient times (Sudrik et al., 2012; Chu, 2001). Fresh juice of betel leaves is used in many ayurvedic preparations (Sharma, 1991). It is used as a traditional remedy of various diseases like bad breath, boils and abscesses, stomach ailments, conjunctivitis, constipation, headache, hysteria, itches, mastitis, mastoiditis, leucorrhoea, otorrhoea, ringworm, swelling of gum, rheumatism, abrasion, cuts and injuries, etc. as folk medicine (Chopra et al., 1956; Khanra, 1997). The Indian traditional system of medicine has identified the leaves as an agent of wound healing (http//www.globinmed. com/index.php?option =com_content&view=article&id=79090:piperbetle-linn&catid=199&Itemid=139; Rahman, 2009) digestive, pancreatic lipase stimulant (Prabhu et al., 1995; Mula et al., 2008; Santhanam and Nagarajan, 1990; Chatterjee and Pakrashi, 1995; Deshpande et al., 1970; Rawat et al., 1989) and deworming. Leaves are considered to use in treating bronchitis, dyspnea (Mula et al., 2008). Leaves were often heated and applied to the chest to relieve cough and asthma (Kumar, 1999). The fruit of betel vine employed with honey for cough treatment (Chandra et al., 1987). It is used as eye drops for eye injury/infection, as a baby lotion for the new born, for coughs, asthma, and constipation; and to arrest milk secretion (Vossen and Wessel, 2000). The fresh betel leaves possess antimicrobial, anti-ringworm, antifungal, antiseptic and anti-helminthic effects (Sarkar et al., 2000). Essential oil of leaves is used for the treatment of respiratory catarrhs and antiseptic (Chandra et al., 1987; Amalia et al., 2008). The volatile oil from the leaves was used as gargle or inhalation in diphtheria and respiratory catarah (Prokopczyk et al., 1991). Leaf extract is reported to inhibit male reproductive competence (Ratnasooriya et al., 1990; 1996; 1997). The leaves possess antifertility on male rats ((Ratnasooriya et al., 1996) and antimotility effects on human spermatozoa ((Ratnasooriya et al., 1990). The leaves were rolled up, covered with oil as a suppository and as purgative in newborns. Not only leaf, the root is known to use as long lasting female oral contraceptive (Agarwal et al., 2012; Das and Patent, 1976) in folk medicine. The piper betel showed hypotensive, cardio tonic, smooth and skeletal muscles relaxant actions (Evans et al., 1984; Bangar et al., 1966; Ali and Mehta, 1970). In spite of these, the betel leaves really works as a cheap, natural and easily available appetizer, digestive, mild stimulant, aphrodisiac and refreshing mastication. The leaves were chewed by singers to improve their voice (Usmanghani et al., 1997). Chewing of betel leaves produces a sense of well-being and increased alertness, sweating, salivation, hot sensation and energetic feeling. The users believe that chewing the leaf improves their efficiency and stamina (Usmanghani et al., 1997).

Besides, the betel vine is commonly used as a folk medicine to treat the following health problems:

Obstructed urination: Betel leaf juice has diuretic properties. A mixture of its juice, diluted milk and slight sugar help in easing the passage of urine.

Neural weakness: It is useful in the treatment of neural disorders. The juice of a few betel leaves, with a teaspoon of honey, serves as a good tonic. A teaspoon of this mixture can be taken twice a day.

Headaches: The leaf of the betel vine credited with analgesic and cooling properties. It is used to relieve intense headaches.

Respiratory disorders: Betel leaves can be applied in pulmonary afflictions suffered in childhood and old age. It, soaked in mustard oil and warmed, can be applied to the chest to relieve a cough or breathing problems.

Constipation: A suppository made of the stalk of betel leaf dipped in castor oil can be carefully introduced in the rectum of children to instantly relieve constipation.

Sore throats: In treating sore throat, the local use of betel vine leaf is effective. The crushed fruit or berry mixing with honey can be applied to relieve an irritating cough.

Wounds: Betel leaves have wounds healing properties. The juice of leaves can be applied on the wound. Then, a leaf of betel vine should be wrapped over it and bandaged. The wound may heal with just a single application.

Boils: It also has an effective boils remedial property. A leaf is softened by gentle warm, and then coated with a layer of castor oil. The oiled leaf is covered over the inflamed part. This leaf has to be replaced, every few hours. After a few applications, the boil will rupture draining out all the purulent matter. This can be applied at night and removed in the morning.

The problem of breast milk secretion: The smear of leaves with oil can be used to promote the secretion of milk when applied to the breasts during lactation.

2.5.2. Biological Activity

The betel leaves possess a number of bioactive compounds. Therefore, it has several medicinal importances on various microorganisms as well as human. The detail biological activities of the betel vine have been described as follows:

Antimicrobial Activity

The betel leaf possess antimicrobial activity against a broad spectrum of microorganisms (Jesonbabu et al., 2012) due to having diverse antimicrobial compound and secretory antibodies (slg A), which offer protection against microbial proliferation in the mouth so that tooth and gum decay is kept under check (Kumar et al., 2010; Chandra and Sagar, 2004). It has the antimicrobial activity against *Streptococcus pyrogen, Staphylococcus aureus, Proteus vulgaris, E. coli, Pseudomonas aeruginosa,* etc., as well as bactericidal activity against the pathogenic bacteria such as *Enterocococcus faecalis, C. koseri, C. fruendi, Klebsiella pnemoniae,* etc in the urinary tract (Agarwal and Singh, 2012; Chakraborty and Shah, 2011). The leaf contains large quantity of bioactive molecule sterol, which is supposed to be responsible for anti-bacterial activity. The sterol of leaf extracts may interact with the bacterial cell wall and membrane leading to alteration in the primary structure of the cell wall, ultimately damage the bacterial cell. The study reported that sterol disrupts the permeability barrier of microbial membrane structures (Chakraborty and Shah, 2011). Gram-positive bacteria were more susceptible to the inhibitory effects of the betel extract because of single layer and lack of natural sieve effect against large molecules, whereas the cell walls of gram-negative bacteria are multi layered with complex structure (Agarwal and Singh, 2012; Scherrer and Gerhardt, 1971). The leaf has also antifungal activity against many fungal infections (Ali et al., 2010), such as dermatophytosis. Dermatophytosis is a disease of the keratinized parts of the body (skin, hair, and nail) caused by a three genera (*Trichophyton, Microsporum,* and *Epidermophyton*) of highly specialized fungi called the Dermatophytes (Trakranrungsie et al., 2006). The chloroform extracts of piper betel shows much more efficiency than that of methanol extract against dermatophytes because, extract contains non-polar components (Sharma et al., 2011).

Gastro-Protective Activity

The leaf extracts protects the gastrointestinal tract by secreting the mucus. Mucus layer is considered to be important in mucosal defenses against endogenous aggressors, e.g., acids, and also as an agent in facilitate the repair process. The hot water extracts of betel leaf is responsible for increasing the mucus secretion of the gastric mucosa, which protects the gastric wall from the induction of gastric lesions or ulcer caused by acid secretion. The leaf extracts has no ability to regulate the acid secretion in the gastric mucosa but helps to increase its mucus content. At a higher dose, the gastro-protective activities of its hot water extract is significantly greater than Misoprostol (Rahmatullah et al., 2009). The extensive research has been proven that anti-oxidants might be the effective mechanism not only in protecting against gastric mucosal injury, but also inhibiting progression of gastric ulceration. Ulceration progression is caused by a free radical-induced chain process. Consequently, it is arrested by radical scavengers and helps in the faster healing process (Majumdar et al., 2003; Arambewela et al., 2004).

Antioxidant Activity

The betel leaf has potential antioxidant properties (Arambewela et al., 2006). It possesses the polyphenols compounds, such as catechol, allylpyrocatecol etc., which is responsible for antioxidant activity. Oxidative damage is an important effect of ionizing radiation on biological membranes. It is a chain reaction (Verma et al., 2010). The free radicals generated from the radiolytic decomposition of water can attack fatty acid chains of membrane lipid. A free radical that has sufficient energy to abstract an allylic hydrogen from the methylene carbon of polyunsaturated fatty acids can initiate the peroxidative process. The catechol and allylpyrocatecol compounds in betel leaf extract inhibit the radiation-induced lipid peroxidation process effectively. This could be attributed to its ability to scavenge free radicals involved in initiation and propagation steps (http://psasir.upm.edu.my/133/). The leaf extracts reduce most of the Fe^{3+} ions and possess strong reductive ability (Manigauha et al., 2009). The extract also has a strong hydroxyl radical and superoxide anion radical scavenging property when compared with different standards such as, ascorbic acid (Rathee et al., 2006; Dasgupta and De, 2004; Pin et al., 2010; Arambewela et al., 2006).

Antidiabetic Activity

The aqueous extracts of betel leaves possess antidiabetic property. Its hypoglycaemic activity was significantly tested in the rat model (Chandra et al., 2011). The leaf extracts markedly reduced the external glucose load. The leaf suspension significantly reduces the blood-glucose level, glycosylated haemoglobin and decreased activities of liver glucose-6-phosphatase and fructose-1, 6- bisphosphatase, whereas liver hexokinase increased in Streptozocin (STZ) diabetic rats. The ability of lowering blood-glucose level of Streptozocin (STZ) induced diabetic rat gives an indication that the extracts have the insulinomimetic activity (Arambewela et al., 2005; Ramji et al., 2002).

Radio Protective Activity

The ethanolic extract of betel leaf has the radioprotective activity. It has been studied using rat liver mitochondria and pBR 322 plasmid DNA. The extract effectively prevented γ-ray induced lipid peroxidation as assessed by measuring thiobarbituric acid reactive substrates, lipid hydroperoxide and conjugated diene. Likewise, it has the capability to prevent radiation-induced DNA damage. The radioprotective activity of betel leaf could be attributed to its hydroxyl and superoxide radicals scavenging property along with its lymphoproliferative activity. The radical scavenging capacity of betel leaf was primarily due to having phenolic compounds, which were isolated and identified as chavibetol and allyl pyrocatechol (Bhattacharya et al., 2005; Zeng et al., 1997).

Effect on the Cardiovascular System/Platelet Inhibition Activity

The betel leaf serves as cardio tonic to strength the heart and regulates irregular heart beat and blood pressure (http://www.ejournalofdentistry.com/ebook/Issue3/data/pages/8.swf.). The heart shape of betel leaf makes it a suitable candidate for heart-related curative properties/medicine (Kumar et al., 2010). Therefore, its name “green heart of nature” carries the significance meaning with its function of heart treatment. It is a feasible substitute for *Digitalis purpurea* (Sharma and Dravyaguna, 1995). The chewing effect of leaf can be found within minutes (Chu, 2001) by exhibiting cardio-acceleration, sweating and salivation. It induces catecholamine secretion from the adrenal cortex contributing to increase in stamina, heart rate, blood pressure and sympathetic neural activity. Platelet hyperactivity is important in the pathogenesis of cardiovascular diseases due to intravascular thrombosis. Piperbetol, ethylpiperbetol, piperol A and piperol B of leaves, selectively inhibits the platelet aggregation induced by platelet activating factor (PAF) (Pisar et al., 2007).

Antifertility Activity

The circulating level of estrogen in the body regulates the structural and functional integrity of reproductive organ. Any small change in estrogen level may lead to alter structural and functional activities of reproductive system (Sharma et al., 2007). The betel leaf extracts may diminish gonadolrophine release through the effect of pituitary-gonadal axis, which in turn reduced reproductive organ weights and estrogen level affecting ovarian cyst. Serum biochemistry revealed that glucose level was declined, but cholesterol and Vitamin C concentrations were increased beyond control value; indicate non-utilization of cholesterol by the system, hence decrease in estrogen level. Thus, antifertility and antiestrogonic effects of the betel vine found in female rats, and these effects were reversible on cessation of treatment (Balamurugan et al., 2009; Priya et al., 2012).

Immunomodulatory Activity

Recently, imbalances of immunological processes have been proved as the cause of many disorders in the human body. Therefore, exploration of newer and safer immunomodulators is necessary. The betel vine has immunomodulatory properties. The methanolic extract of betel leaf has lymphocyte proliferation, interferon-C receptors and nitric oxide production properties. The studies proved the immunosuppressive effect of betel extracts on cellular and humoral immune responses in mice (Kanjwani et al., 2008). Consequently, betel leaf has been recognized as a novel candidate for immunosuppressive activity. The same could be further evaluated for its anticancer activity or as a potential candidate in the treatment of autoimmune disorders such as rheumatoid arthritis, systemic lupus erythomatous or emphysema.

Cholinomimetic Effect

The betel leaf helps to raise body temperature due to cholinergic responses. Aqueous and ethyl acetate extracts of betel vine leaf have cholinergic responses. Both water and ethyl acetate extracts of betel leaf have spasmogenic activity. In isolated rabbit jejunum K^+-induced contraction was inhibited by both extracts, suggesting blockades in the calcium-channel. Thus, leaves contain cholinomimetic and possible calcium-channel antagonist constituents who may provide the basis of several activities shown by this plant (Gilani et al., 2000).

Hepato-Protective Activity

The extracts of betel leaf also play important roles in the process of hepatic protection. Its antihepatotoxic effect was evaluated on ethanol and carbon tetrachloride (CCl_4) induced liver injury in a rat model (Saravanan et al., 2003). Fibrosis and hepatic damages are induced in rats by CCl_4 as revealed by the analysis of histology and the activities of aspartate aminotransferase (AST) and alanine aminotransferase (ALT). The extracts of betel leaf significantly inhibit the elevated activities of AST and ALT and also attenuate total glutathione S-transferase, which led to a rise in antioxidant enzymes such as superoxide dismutase and catalase (Kumar et al., 2010). The histological study shows that the betel leaf extracts protect liver from the damage induced by CCl_4 by decreasing alphasmooth muscle actin (alpha-sma) expression, inducing active-matrix metalloproteinase-2 (MMP2) expression through the Ras/Erk pathway, and inhibiting TIMP2 level that consequently attenuated the fibrosis of liver. These findings support a chemo preventive potential of betel leaf against liver fibrosis (Young et al., 2007).

As an Oral Care Agent

The aqueous extract of the betel leaf is a good oral care agent. The normal oral commensally flora is responsible for causing chronic endogenous infection - dental caries. The dental carious lesion is the result of demineralization of enamel and layer of dentine by acids produced by plaque microorganisms as they metabolize dietary carbohydrates (Bowden, 2000; http://en.wikipedia.org/wiki/Dental_caries; http://www. aquafreshscienceacademy.com/oral-health/enamel-protection/ dentalcavities.html.). The bacteria like, *Streptococcus mutans*. *Streptococci salivarius, Streptococci anginosus, Streptococci mitis, Lactobacillus acidophilus and Staphylococci* sp. probably plays a vital role in dental decay by producing acid in the plaque (http://www.slideshare.net/ElsevierIndia/11sa mple-ch-rajendranindd.). The stickiness of the plaque is caused by dextran, which is produced by the fermentation of dietary sucrose by *Streptococcus mutans*. The plaque bacteria, particularly *Streptococcus mutans*, act on dietary fructose to produce lactic acid, which causes enamel decalcification (at below or above 5.5 pH) (Stoppelaa, 1971). The aqueous extract inhibits the above-mentioned different acid-producing oral pathogens which change the ultra-structure of the enamel and its properties. So, it is a best natural substance and its rating as second most popular daily consummation item in Asia, which contribute the best oral hygiene to oral cavity (Bissa et al., 2007).

Neuropharmacological Profile

Hydro-alcoholic extract of betel leaves exhibited improvement in the discrimination index, potentiating the haloperidol induced catalepsy, reduction in basal as well as amphetamine induced increased locomotors activity and delay in sodium nitrite induced respiratory arrest. These results from a review suggest possible facilitation of cholinergic transmission and inhibition of dopaminergic as well as nor an adrenergic transmission by the extract (Razak and Rahim, 2003; Razak and Rahim, 2007).

Anti-Carcinogenic Activity

It has significant anti-carcinogenic properties used for manufacturing of a promising blood cancer drug. In this context, some disputed reports also claim that chewing of betel leaves excessively may cause oral cancer, especially in the tongue and cheeks. In areas of widely betel chewing habit, the cancer in mouth and lips has been found to be more frequent. Other ill-effects of betel leave chewing are dyspepsia and pyorrhea. Study evaluated an aqueous extract of leaves to cytotoxicity studies on Hep-2 cell line, which concluded that it has potent cytotoxicity and probable anticancer property (Dwivedi et al., 2010). Gundala et al. (2014) proposed that the phenolic compound of betel leaf inhibits prostate cancer through reactive oxygen species (ROS) driven DNA damage and apoptosis.

Figure 22. Betel vine grown in the tub kept for indoor decoration.

2.6. Industrial Value

In spite of medicinal applications, the extracted oil of betel leaves is used as an industrial raw material for manufacturing cosmetics, perfumes, mouth fresheners, tonics, food additives etc. Several value-added products from betel have been formulated and those include betel toothpaste, mouthwash, shampoo, face cream, instant betel quid and pellets, etc.

Figure 23. The betel vines covers the soil surface of the cultivated area and thus conserve the soil by preventing soil erosion.

2.7. Aesthetic Use

The betel vine plant is grown in garden and in the tub for beautification and decoration because of its glossy and characteristic heart-shaped leaf (Figure 22). It is used as decorative or artistic purposes in different cultural occasions and functions of some countries like India, Nepal, Bangladesh and Vietnam (Figure 22).

2.8. Environmental Importance

The betel vine also may play an important role in conserving the environment. Although it climbs up using support tree/sticks as high as 10 to 15 ft, sometimes the plant can also grow as a ground cover (Figure 23) and it has a growth pattern similar to that of the pepper plant. Therefore, it directly and indirectly conserves the soil by preventing the soil erosion in cultivated areas. Its adventitious root could be used as a biological tool to bioremediate the pollutants from contaminated soil by uptaking hazardous pollutants. Because, it is known that the stem having adventitious roots is folded in the regular lowering process of betel vine cultivation and substantial amount of fertile soil is required to cover over the folded stem. It is a routine management practice in a closed and organized cultivation process to supply the nutrients through the soil. Such, plants can uptake and remove pollutants from soil. A substantial research work is required in this regard. Besides, its fast-growth property is associated with the utilization of a substantial amount of green house gas carbon dioxide from atmosphere. Such, it may also play a vital role in controlling the air pollution. In this sense, it has high rate of carbon sequestration potentiality as well as valuable plant leaf biomass production. In spite of above various applied importances, the excess waste leaf and stem may be used for producing bio-compost manure for applying as bio-fertilizer in agriculture field.

Table 10. Patents filed and awarded for various activities of *Piper betle*

Patent filed for activity	Patent title and abstract	Place of filing	Filing country/year	Grant date/patent number
Antileishmanial	Antileishmanial activity of betel-leaf extract. (This invention relates to the method of treating VL/kala-azar by administering an effective amount of betel-leaf extract or lyophilized extract together with or associated with an additive and a composition comprising betel-leaf extract with a pharmaceutically acceptable additive.)	India (CSIR/ IICB, Kolkata)	World/2000; USA/2001, 2002; China/2003; India/2003	WO/20002/045731 USA/2003/6610332
Anticancer	Antimonocytic activity of extracts of PBL. (This invention relates to anti-monocytic activity of betel-leaf extracts and suggest its use in the treatment of myeloid leukaemia in animals and human beings.)	India (CSIR/ IICB, Kolkata)	World/2000; USA/2001; Australia/2003; GB/2003; Japan/2003; China/2003; India/2003; Denmark/2003	GB/2004
Immunomodulatory	Use of betel-leaf extract to induce IFN-gamma production from human peripheral blood T-cells and as a Th1-type immunomodulator	India (CSIR/IICB, Kolkata)	World/2000; USA/2002, 2005; India/2003; China/2003; Japan/2003 Denmark/2003	WO/2002/049655 USA/2003/6531166
Anti-5	Herbal formulation of a combination of PB and	India	USA/2001;	USA/2004/6773728

Patent filed for activity	Patent title and abstract	Place of filing	Filing country/year	Grant date/patent number
lipoxygenase	*Murrya koenigii* extracts for blocking 5-lipoxygenase activity leading to the inhibition of leukotriene synthesis, suppression of interleukin-4 production, and enhancement of gamma interferon release	(CSIR/IICB, Kolkata)	World/2002; Australia/2004; Europe/2004; Japan/2004; China/2004	Europe/2005
Anticancer	A herbal composition for treating CD33+ acute and chronic myeloid leukaemia and a method thereof.	India (CSIR/IICB, Kolkata)	USA/2004; USA/2002, 2003	USA/2007/7306817 USA/2005/6852344 2005/6967034
Antiwart	Method for instantaneous removal of warts and moles.	USA (Deerfield, IL)	USA/2001	USA/2001/6312735
Dye	Herbal dye and process of preparation thereof. (The present invention provides a herbal black dye from natural materials comprising *Juglans regia*, *Indigofera tinctoria*, *Terminalia chebula*, *Acacia accocina*, *Lawsonia inermis*, *Trigonella foenum-graecum*, *Sapindus mukorossi*, *Eclipta alba*, *Embelica officinalis*, *Acacia catechu* and PB. The dye derived is safe, non-toxic, antiallergic, antidandruff and free from toxic symptoms like itching.)	India (CSIR/NBRI, Lucknow)	USA/2004	USA/2007/7186279

Table 10. (Continued).

Patent filed for activity	Patent title and abstract	Place of filing	Filing country/year	Grant date/patent number
Bronchial disorders	Herbal composition of a blend of active components prepared. (This invention relates to a herbal composition for the treatment and as a remedy for bronchial respiratory difficulties. More particularly, this invention describes the process of separation, physico-chemical characterization and biological response evaluation of active components obtained from extracts of any plant parts, including leaves, barks, roots and seeds of *M. koenigii* and PB plants in order to establish their role in the treatment and as a remedy for bronchial respiratory troubles.)	India (CSIR/IICB, Kolkata)	USA/2001	USA/2004/6773728
Anti-inflammatory	Analgesic and refreshing herbal composition and a process for preparing the same. (This invention provides an analgesic and refreshing herbal composition useful as dentrifrices; composition comprising 50–60% wt of betle extract; from PBL.)	India (CSIR/CIMAP, Lucknow)	USA/2001	USA/2003/6531115

On account of above, it is obvious that several scientists have carried out a number of quality researches and discovered various significant properties of the betel vine. In this respect, many outstanding research findings have also obtained patents as listed in Table 10 (Kumar et al., 2010).

REFERENCES

Agarwal, T. and Singh, R. (2012). Evaluation of Antimicrobial Activity of Piper betel cultivars. *Novus International Journal of Pharmaceutical Technology,* 1(1): 50-58.

Agarwal, T., Singh, R., Shukla, A. D., Waris, I. and Gujrati, A. (2012). Comparative analysis of antibacterial activity of four *Piper betel* varieties. *Advances in Applied Science Research*, 3(2): 698-705.

Ali, I., Khan, F. G., Suri, K. A., Gupta, B. D., Satti, N. K., Dutt, P., Afrin, F., Qazi, G. N. and Khan, I. A. (2010). In vitro antifungal activity of hydroxychavicol isolated from *Piper betle* L. *Annals of Clinical Microbiology and Antimicrobials,* 9(7): 1-9.

Ali, S. M. and Mehta, R. K. (1970). Preliminary pharmacological and anthelmintic studies of the essential oil of *Piper betle. Ind. J. Pharm.*, 32: pp. 132.

Amalia, H., Sitompul, R., Hutauruk, J. and Andrianjah Mun'im A. (2008). Effectiveness of Piper betle leaf infusion as a palpebral skin antiseptic. *Universa Medicina,* 28(2): 83-91.

Arambewela, L. S. R., Arawwawala, L. D. A. M. and Ratnasooriya, W. D. (2004). Gastroprotective activities of Sri Lankan Piper betle leaf extracts in rats. SLAAS, *60th Annual Session*, pp. 117.

Arambewela, L. S., Arawwawala, L. D. and Ratnasooriya, W. D. (2005). Antidiabetic activity of aqueous and ethanolic extract of *Piper betle. J. Ethanopharmacol.,* 102(2): 239-245.

Arambewela, L., Arawwawala, M. and Rajapaksa, D. (2006). *Piper betle*: a potential natural antioxidant. *International Journal of Food Science andTechnology,* 41 (1): 10-14.

Balamurugan, K., Kalaichelvan, V. K, Anuradha, G., Madhana, G. K., Meganathan, M. and Manavalan, R. (2009). Antifertility activity of methanolic extract of Saccharum officinarum Linn. (Leaves) on Female Albino Rats. *International Journal of Pharm. Tech. Research*, 1.1(4): 1621-1624.

Bangar, G. P., Rao, R. E. and Varma, K. C. (1966). Antimicrobial activity of leaves and oil of *Piper betle. Ind. J. Pharm.*, 28: pp. 327.

Bhakta, B., Bhakta, U. Ghanti, S., Jana, B. B., Ghanti, P., Yamane, S., Bhakta, J. N. (2016) Betelvine: A green gold for sustainable economic upliftment in rural India. *Int. J. Env. & Tech. Sc.* 3:76-82.

Bhattacharya, S., Subramanian, M., Roychowdhudy, S., Bauri, A. K., Kamat, J. P. and Chattopadhyay, S. (2005). Radio protective property of the ethanolic extract of Piper betle leaf. *J. Radiat. Res.*, 46(2): 165-171.

Bissa, S., Songara, D. and Bohra, A. (2007). Traditions in oral hygiene: Chewing of betel (*Piper betle* L.) leaves. *Current Science*, 92(1): 26-28.

Bowden, G. H. W. (2000). The Microbial Ecology of Dental Caries. *Microbial Ecology in Health and Disease*, 12: 138–148.

Chakraborty, D. and Shah, B. (2011). Antimicrobial, antioxidative and antihemolytic activity of Piper betel leaf extracts. *Int. J. Pharm. Pharm. Sci.*, 3(3): 192199.

Chandra, G. and Sagar, R. L. (2004). Harvesting green gold: Cultivation of betelvine in Sundarban. *Indian Farmers Digest*, 37(3): 5-13.

Chandra, T., Sadique, J. and Somasundaram, S. (1987). Effect of *Elipta alba* on inflammation and liver injury. *Fitoterapia*, 58: 23-31.

Chandra, V., Tripathi, S., Verma, N.K., Singh, D.P., Chaudhary, S.K. and Roshan, A. (2011). *Piper betel*: phytochemistry, traditional use & pharmacological activity-a review. *IJPRD*, 4(4): 216-223.

Chatterjee, A. and Pakrashi, S. C. (1995). Treatise of Indian Medicinal Plants. CSIR Publication, New Delhi, India, pp. 26.

Chopra, R. N., Nayar, S. L. and Chopra, I. C. (1956). Glossary of Indian Medicinal Plants. CSIR, New Delhi, India, pp. 194.

Chu, N. S. (2001). Effects of Betel Chewing on the Central and Autonomic Nervous Systems. *J Biomed Sci.,* 8: 229-236.

Chu, N. S. (2001). Effects of betel chewing on the central and autonomic nervous systems. *J. Biomed. Sci.,* 8: 229-236.

CSIR (Council of Scientific and Industrial Research) (1969). The Wealth of India. 8: 84-94. CSIR, New Delhi, India.

CSIR (Council of Scientific and Industrial Research) 1992. The wealth of India, The dictionary of Indian raw materials and industrial products. Raw material, New Delhi, (Publication and information directorate,), pp. 5.

Das, P. C. and Patent, G. B. (1976). *Chemical Abstracts*, 86: 21786.

Dasgupta, N. and De, B., (2004). Antioxidant activity of Piper betle L. leaf extract in vitro. *Food Chemistry,* 88(2): 219-224.

Deshpande, S. M., Upadhyay, R. R. and Singh, R. P., (1970). Chemical study of Piper betel leaves. *Current Sci.,* 39: 372.

Dwivedi, B. K., Kumar, S., Nayak, C. and Mehta, B. K. (2010). Gas chromatography mass spectrometry (GCMS) analysis of the hexane and benzeneextracts of the *Piper betle* (leaf stalk) (Family: Piperaceae) from India. *J Med Plants Res.* 4(21): 2252-2255.

Evans, P. H., Bowers, W. S. and Funk, E. J. (1984). Identification of Fungicidal and Nematocidal Components in the Leaves of *Piper betel* (Piperaceae). *Journal of Food Agricultural andChemistry,* 32: 1254-1256.

Gilani, A. H., Khurram, Aziz. N., I. M., Rao, Z. A. and Ali, N. K. (2000). The presence of Cholinomimetic and *Piper betle* Linn. *Phytotherapy Research,* 14(6): 436-442.

Gopalan, C., Ramasastri, B. V. and Balasubramanian, S. C. (1984). Nutritive value of Indian foods. National Institute of Nutrition (ICMR), Hyderabad, India, pp. 108.

Guha, P. (1997). Paan Theke Kutir Silpa Sambhabana" (In Bengali). In: Krishi, Khadya-O- Gramin Bikash Mela, Agricultural and Food Engineering Department, IIT, Kharagpur, India, pp. 15-19. [Guha, P. (1997). Exploring Betel Leaves for Cottage Industry. In: Festival on

Agriculture, food, and development of village, Agricultural and Food Engineering Department, IIT, Kharagpur, India, pp. 15-19.]

Guha, P. (2006). Betel leaf: The neglected green gold of India. *Journal of Human Ecology,* 19(2): 87-93.

Gundala, S. R., Yang, C., Mukkavilli, R., Paranjpe, R., Brahmbhatt, M., Pannu, V., et al. (2014). Hydroxychavicol, a betel leaf component, inhibits prostate cancer through ROS-driven DNA damage and apoptosis. *Toxicol Appl Pharm,* 280(1): 86-96.

http://en.wikipedia.org/wiki/Dental_caries. Dental caries. (22 November, 2012).

http://psasir.upm.edu.my/133/Antimicrobial Activity of Psidium Guajava and *Piper betle* Extracts on Selected Foodborne Bacteria. (12 May, 2006).

http://www.aquafreshscienceacademy.com/oral-health/enamel-protection/dentalca vities. html. Dental-cavities.

http://www.ejournalofdentistry.com/ebook/Issue3/data/pages/8.swf. Areca nut: To chew or not to chew? (22 October, 2012).

http://www.globinmed.com/index.php?option=com_content&view=article &id=79090:piperbetle-linn&catid=199&Itemid=139. *Piper betle* Linn. 2 December, 2012.

http://www.slideshare.net/ElsevierIndia/11sa mple-ch-rajendranindd. (22 April, 2012).

Jana, B. L. (1996). Improved technology for betel leaf cultivation. A paper presented in the *"Seminar-cum-Workshop on Betel leaf Marketing",* held on June 5-6, 1996 at state cashew nut farm, Directorate of Agricultural Marketing, Digha, Midnapur, India.

Jesonbabu, J., Spandana, N. and Lakshmi, K. A. (2012). *Invitro* antimicrobial potentialities of chloroform extracts of Ethanomedicinal plant against clinically isolated human pathogens. *Int. J.Pharm. Pharm. Sci.,* 2012: 4(3): 624-626.

Kanjwani, D. G., Marathe, T. P., Chiplunkar, S. V. and Sathaye, S. S. (2008). Evaluation of immunomodulatory activity of methanolic extract of *Piper betel. Scand J. Immunol.,* 67(6): 589-593.

Khanra, S. (1997). Paan Vittik Silpakendra. *Nabanna Bharati,* 30(2): 169. (In Bengali). [Khanra, S. (1997). Betel Leaf Based Industry. *Nabanna Bharati,* 30(2): 169.]

Khoshoo, T. N. (1981). Welcome address, In: *Proc. of Group Discussion on Improvement of Betelvine Cultivation.* Khanduja, S. D. and Balasubrahmanyam V. R. (Eds.), National Botanical Research Institute, Lucknow, India, pp. xvii-xx.

Kumar, N. (1999). Betalvine (*Piper betle* L.) cultivation: A unique case of plant establishment under anthropogenically regulated microclimatic conditions. *Indian journal of History of science,* 34(1): 19-32.

Kumar, N., Misra, P., Dube, A., Bhattacharya, S., Dikshit, M. and Ranade, S (2010). *Piper betle* Linn. a maligned Pan-Asiatic plant with an array of pharmacological activities and prospects for drug discovery. *Current science,* 99(7): 922-932.

Majumdar, B., Ray, C. S. G., Ray, A. and Bandyopadhyay, S. K. (2003). Effect of ethanol extract of *Piper betle* Linn leaf on healing of NSAID induced experimental ulcer a novel role of free radical scavenging action. *Indian J Exp Biol.*, 41(4): 311-315.

Manigauha, A., Ali, H. and Maheshwari, M. U. (2009). Antioxidant activity of ethanolic extract of Piper betel leaves. *Journal of Pharmacy Research,* 2(3):194-95.

Mula, S., Banerjee, D., Patro, B. S., Barik, A., Bandopadhayay, S. K. and Chattopadhyay S. (2008). Inhibitory property of the Piper betle phenolics against photosensitization induced biological damages. *Bioorganic and medicinal chemistry,* 16: 2932-2938.

Pin, K.Y., Chuah, A. L., Rashih, A. A., Mazura, M. P., Fadzureena J., Vimala, S. and Rasadah, M. A. (2010). Antioxidant and anti-inflammatory activities of Extracts of betel leaves (*Piper betle*) from solvents with different polarities. *Journal of Tropical Forest Science,* 22(4) 448-455.

Pisar, M. Md., Hashim, N., Ali, R. M. and Sui, K. L. (2007). Evaluation of *Piper betle* on Platelet Activating Factor (PAF) Receptor Binding Activities. *Malaysian Journal of Science*, 26(1): 79-83.

Prabhu, M. S., Platel, K., Saraswathi, G. and Srinivasan, K. (1995). Effect of orally administered betel leaf (*Piper betle* Linn.) on digestive enzymes of pancreas and intestinal mucosa and on bile production in rats. *Indian J Exp Biol.,* 33(10): 752-756.

Priya, G., Saravanan, K. and Renuka, C. (2012). Medicinal plants with potential antifertility activity- A review of sixteen years of herbal medicine research (1994-2010). *International Journal of PharmTech Research,* 4(1): 481-494.

Prokopczyk, B., Rivenson, A. and Hoffmann, D. (1991). A study of betel quid carcinogenesis IX. Comparative carcinogenicity of 3-(methylnitrosamino) propionitrile and 4- (methylnitrosamino)-1-(3-pyridyl)-1-butanone upon local application to mouse skin and rat oral mucosa. *Cancer Letters,* 60(2): 153-157.

Rahman, S. A. (2009). Anti-ulcer effects of Piper betel, Solanum nigrum and Zingibercassumunar on ulceration induced by selected ulcerogens in rats. Master's thesis, University Putra Malaysia, pp. 4.

Rahmatullah, M., Mukti, I. J., Haque, A. K. M. F., Mollik, Md. A. H., Kanta, P., Jahan, R., Chowdhury, M. H. and Taufiq, R., (2009). An Ethnobotanical Survey and Pharmacological Evaluation of Medicinal Plants used by the Garo Tribal Community living in Netrakona district, Bangladesh. *Advances inNatural and Applied Sciences,* 3(3): 402-418.

Ramji, N., Ramji, N., Iyer, R. and Chandrasekaran, S. (2002) Phenolic antibacterial from *Piper betle* in the prevention of halitosis. *J. Ethonpharmacology*, 83: 149-152.

Rathee, J. S., Patro, B. S., Mula, S., Gamre, S. and Chattopadhyay, S., (2006). Antioxidant Activity of Piper betel Leaf Extract and Its Constituents. *J. Agric.Food Chem.*, 54(24): 90460-9054.

Ratnasooriya, W. D. and Premakumara, G. A. S. (1997). *Piper betle* leaves reversibly inhibits fertility of male rats. *Vidyodaya Journal of Science*, 7: 15-21.

Ratnasooriya, W. D. and Premakumara, G. A. S., *Piper betle* leaves impairs masculine sexual behaviour of rats. *Medical Science Research,* 24: 303-306.

Ratnasooriya, W. D., Jayawardena, K. G. I. and Premkumara G. A. S. (1990). Antimotility effect of *Piper betle* (L) leaf extract on washed human spermatozoa. *Journal of National Science Council of Srilanka*, 18: 53-60.

Rawat, A. K. S., Tripathi, R. D., Khan, A. J. and Balasubrahmanyam, V. R., (1989). Essential oil components as markers for identification of *Piper betle* L. Cultivars. *Biochem. Syst. Ecol.*, 17: 38-55.

Razak, F. A. and Rahim, Z. H. A. (2003). The anti-adherence effect of *Piper betle* and *Psidium guajava* extracts on the adhesion of early settlers in dental plaque to saliva-coated glass surfaces. *Journal of Oral Science,* 45(4): 201-206.

Rooney, D. F. (1993). Betel Chewing Traditions in South-East Asia. Kuala Lumpur, Oxford University Press.

Samanta, C. (1994). Paan chaser samasyabali-o-samadhan: Ekti samikkha (In Bengali) [A Report on the Problems and Solutions of Betel Vine Cultivation]. In: Adhikari, H. R., Salt Lake City, Kolkata, India.

Santhanam, G. and Nagarajan, S. (1990). "Wound healing activity of Curcuma aromatica and Piper betle. *Fitoterapia,* 61: 458-459.

Saravanan, R., Rajendra, N., Prasad, N. R. and Pugalendi, K. V. (2003). Effect of *Piper betle* leaf extract on alcohol toxicity in the rat brain. *J. Med. Food,* 6(3): 261-265.

Sarkar, M., Gangopadhyay, P. and Basak, B. (2000). The reversible antifertility activity of Piper betel Linn. on Swiss albino male mice. *Contraception,* 62: 271-274.

Scherrer, R. and Gerhardt, P. (1971). Molecular sieving by the *Bacillus megaterium* cell wall and protoplast. *J. Bacteriol,* 107: 718-735.

Sharma, J. D., Sharma, L. and Yadav, P. (2007). Antifertility Efficacy of *Piper betle* Linn. (Petiole) on Female Albino Rats. *Asian J. Exp. Sci.,* 21(1): 145-150.

Sharma, K. K., Saikia, R., Kotoky, J., Kalita, J. C. and Das, J. (2011) Evaluation of Antidermatophytic activity of *Piper betle*, *Allamanda cathertica* and their combination: An in vitro and in vivo stud. *International Journal of PharmTech. Research,* 3(2): 644-651.

Sharma, M. L., Rawat, A. K. S., Khanna, R. K., Chowdhury, A. R. and Raina, R. M. (1996). Flavour characteristics of betel leaves. *Euro cosmetics*, 5: 22-24.

Sharma, P. V. and Dravyaguna, V. (1995). Vegetable Drugs, Chaukhambha Sanskrit Bhawan, Varanasi, Vol. 2.

Sharma, R. N. (1991). Contribution of Ayurveda towards management of bronchial asthma. *Sachitra Ayurveda*, 44: 349-352.

Stoppelaa, J. D. De. (1971) Decreased carcinogenicity of a mutant of *Streptococcus mutans. Archs oral Bid.*, 16: 971-975.

Sudrik, S., Fegade S. and Shinde M. (2012). Anthelmintic Activity of *Piper Betle* Linn, (Paan/Nagavalli) Aqueous Extract. *Research Journal of Pharmaceutical, Biological and ChemicalSciences,* 3(4): 467-470.

Trakranrungsie, N., Chatchawanchonteera, A. and Khunkitti, W. (2006). Antidermatophytic Activity of *Piper betle* Cream. *Thai J Pharmacol.,* 28(3): 16-20.

Usmanghani, K., Saeed, A. and Alam, M.T. (1997). *Piper betle* In, Indusyunic Medicine, University of Karachi Press, Karachi, pp. 340-341.

Verma, S., Gupta, M. L., Dutta A., Sankhwar, S., Shukla, S. K. and Flora, S. J. (2010). Modulation of ionizing radiation induced oxidative imbalance by semi-fractionated extract of *Piper betle*: an in vitro and in vivo assessment. *Oxid. Med. Cell Longev.*, 3(1): 44-52.

Vossen, H. A. M. V. and Wessel M. (2000). Plant Resources of South-East Asia Stimulants. Backhuys Publisher, Netherlands, 102-106.

Young, S. C., Wang, C. J., Lin, J. J., Peng, P. L., Hsu, J. L. and Chou, F. P. (2007). Protection effect of piper betel leaf extract against carbon tetrachloride-induced liver fibrosis in rats. *Arch Toxicol.*, 81(1): 45-55.

Zeng, H. W., Jiang, Y. Y., Cai, D. G., Bian, J., Long, K. and Chen, Z. L. (1997). Piperbetol, methylpiperbetol, piperol A and piperol B a new series of highly specific PAF receptor antagonists from Piper betel. *Planta Med.*, 63: 296-298.

Chapter 7

CULTIVATION TECHNOLOGY

1. INTRODUCTION

The betel vine is one of the most important cash crops in many Asian countries, especially in India, Bangladesh, Pakistan, Srilanka, etc. The cultivation of betel vine has also been recognized as an age-old practice as that of betel leaf chewing habit in several South East Asian countries. It has naturally been grown as a wild vine in the natural environment from the dawn of civilization, gotten more interest and domesticated subsequently while human explores its multipurpose significance from ancient times. Presently, it is extensively, intensively and systematically cultivated in the garden or the constructed enclosed or semi-enclosed farming methods as "green gold" or "green heart" of the nature for huge production to meet the demand of present market. The special criterion of betel vine crop is perennial, once sown the betel vine, reaps the leaves for next 8 to 10 years. It has specific cultivation process varied in different areas within a country and from country to country with environmental conditions. Basic technology followed in the betel vine farming process is same that can simply be represented in the flow chart as below (Figure 24). It illustrates a primary concept at a glance in this regard.

2. Types of Cultivation

Since, the betel plant is a creeper; it needs a compatible tree (i.e., support tree) or a long pole for support. Therefore, cultivation techniques depend on the types of supports (natural and artificial) and other various conditions. Generally, on the basis of cultivation practices, it can be divided into following two categories (Kumar, 1999; Haider et al., 2013):

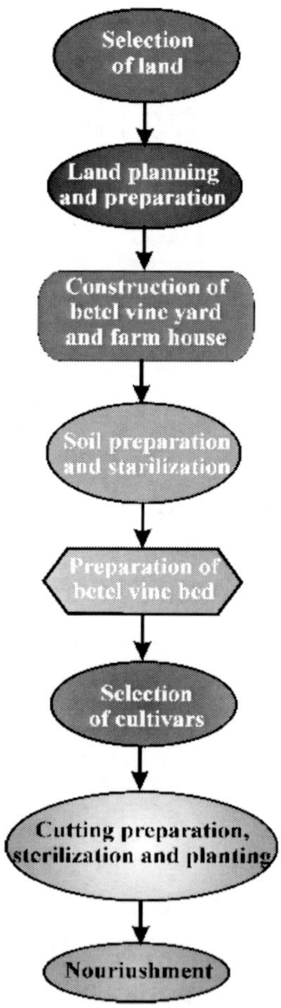

Figure 24. Flow chart representing the preliminary concept of betel vine farming technology.

1. Open or natural cultivation (i.e., bamboo pole or tree based betel-vine cultivation in the open system without enclosure)
2. Closed or controlled cultivation (i.e., betel-vine cultivation within the enclosure system)

In India, both "open or natural cultivation" and "closed or controlled cultivation" technologies of betel vine farming are practiced in different states (Figure 25). Additionally, it is also necessary to mention herein that the cultivation is well spread out in the rest parts of the country with distinctly different methods, specifically for enhancing the crop production.

Figure 25. Major betel vine farming areas in India (C = closed or controlled cultivation and O = open or natural cultivation).

(a) (b)

Figure 26. Tree based betel-leaf cultivation in open place (a and b). The betel vine is growing on the support tree in natural environment.

3. Open or Natural Cultivation

Open or natural cultivation includes the growing of the betel vine on the compatible support trees under natural condition generally in the forests (Figure 26 a,b). It is called as open or natural cultivation, because these cultivation conditions are not controlled as that of the closed or controlled betel vine cultivation process followed. It can be considered as an important beneficial agroforestry. This type of cultivation is found in hilly districts of Eastern part, Andaman and Nikobar, Western ghats and some other parts of India and Bangladesh (Haider et al., 2013). At present, more than 80% Khasia ethnic communities in hilly districts of Eastern part of India are dependent on this betel vine cultivation, and hence it has become their main source of livelihood (Nath et al., 2003; Fila et al., 2006; Alam and Mahiuddin, 1995; Costa and Datta, 2007). This cultivation technique involves the following stepwise process (Kumar, 1999; Haider et al., 2013):

Figure 27. The view of open betel vine farm. The betel vine planted with compatible plants for tree based betel vine farming in open or natural cultivation. An expert person demonstrating the process of a tree based open betel vine cultivation in front of some trainees.

3.1. Establishment of Farm

3.1.1. Selection of Planting Site

The shady forest area with sufficient supporting plants, high humidity, rich soil moister and well drained loamy and sandy loam soil is selected for betel vine cultivation (Figure 27). The supporting plants may be natural or are planted before 4 to 5 years. Generally, the betel vine is cultivated near around the support plant when the support sapling attains 3 to 4 m in height. The different types of timber plants are also grown and employed as support trees in the betel vine cultivated field, such as- *Artocarpus chama, Vitex sp. Syzigium cumini, Toona ciliate, Schima wallichii, Bombax insigne, Anthocephalus chinensis, Ficus* sp., *Syzygium fruticosum, Lannea coromandelica, Artocarpus lacucha, Artocarpus heterophyllus,* etc. (Haider et al., 2013). The following table is the list of various trees found as support plants in the Khasia betel-vine farming practice (Table 11).

In some cases of open or natural betel vine farming, the field is prepared and beds of 2 m wide are formed to a convenient length. Drainage trenches of 0.5 m width by 0.5 m depth are allowed in between two adjoining beds. The seeds of the living support trees, i.e., Agathi (*Sesbania grandiflora*) or Areca tree (*Areca catechu*), etc. are planted in long rows.

In some cases, banana suckers are planted at the edges of the beds, which are used, for tying the vines on the living support and for packing the betel leaf. When the support trees (Agathi or Areca plants) reach 4 m height, they are topped off for maintaining the height. Generally, the crop is planted in two rows in beds of 180 cm width near the support trees with a spacing of about 45 cm between two betel vine plants in the row.

Table 11. The list of various support trees for growing the betel vine in the Khasia betel vine farm of open or natural cultivation (Haider et al., 2013)

Scientific Name	Family
Anacardium Occidentale L.	Anacardiaceae
Holigarna caustica (Dennst.) Oken	-
Lannea coromandelica (Houtt.) Merr.	-
Mangifera sylvatica Roxb.	-
M. Indica L.	-
Spondias pinnata L. f. Kurz	-
Alstonia scholaris (L.) R. Br.	Apocynaceae
Ilex godajam Coleb.	Aquifoliaceae
Caryota urens L.	Arecaceae
Haplophragmaa adenophyllum (wwll.) Dop.	Bignoniaceae
Oroxylum indicum (L.) Vent.	-
Stereospermum personstum (Hassk.) Charttejee	-
Bombax insigne Wall.	Bombacaceae
Protium serratum (Wall. ex Colebr.) Engl.	-
Cassia nodosa Buch.- Ham. Ex Roxb.	Caesalpiniaceae
Garcinia cowa Roxb.	Clusiaceae
G. xanthochymus Hook. F.	-
Terminalia bellirica (Gaertn.) Roxb.	Combretaceae
T. catappa L.	-
T. chebula (Gaertn.) Retz.	-
Dillenia pentagyna Roxb.	Dilleniaceae
Dipterocarpus turbinatus Gaertn.	Dipterocarpaceae
Elaeocarpus varruna Ham.	Elaeocarpaceae
Aporusa dioica (Roxb.)Muell	Euphorbiaceae
Gelonium multiflorum A. Juss.	-
Glochidium sphaerogynum Kurz.	-

Scientific Name	**Family**
Mallotus philippensis Muell. Arg.	-
M. albus Muell. Arg.	-
Macaranga denticulate (Bl.) Muell. Arg.	-
Phyllanthus emblica L.	-
Sapium baccatum Roxb.	-
Erythrina Variegata L. var. orientalis (L.) Meer.	Fabaceae
Castanopsis tribuloides A. DC.	Fabaceae
C. indica (Roxb.) A. DC.	-
Alseodaphne petiolaris Hook.f.	Lauraceae
Cinnamomum cecidodaphne Meissn.	-
Litsea monopetala (Roxb.) Pers.	-
L. glutinosa (Lour) C. B. Rob.	Lecythidaceae
Lagerstroemia parviflora Roxb.	Lythraceae
Amoora wallichii King.	Meliaceae
Aphanamixis polystachya (Wall.) Parker	-
Azadirachta indica A. Juss.	-
Dysoxylum binectariferum Hook. f. exBedd.	-
Toona ciliata M. J. Roem	-
Albizia procera (Roxb.) Benth.	Mimosaceae
A. chinensis (Osbeck) Merr.	-
Artocarpus chama Hamilton	Moraceae
A. lacucha Buch. – Ham.	-
A. heterophyllus Lamk.	-
Ficus auriculata Lour.	-
F. hispida L.	-
F. lepidosa Wallich	-
F. racemosa L.	-
F. clavata Wallich	-
Streblus asper Lour.	-
Ardisia khasiana Clark var. thomsoni Clark	Myrsinaceae
Maesa ramentaceae A. DC.	-
Syzygium cumini (L.) Skeel	Myrtaceae
S. fruticosum (Roxb.) Balak.	-
S. grandis (Wt.) Wallich	-
Xanthophyllum flavescens Roxb.	Polygalaceae
Gardenia coronaria Ham.	Ruaceae
Hymenodiction orixensis (Roxb.) Mabberley	-
M. parviflora (Roxb.) Korth.	-
Mitragyna rotundifolia (Roxb.) O. Ktze.	-

Table 11. (Continued)

Scientific Name	Family
Randia dumetorum Lamk.	-
Tricalysia singularis k. Schum.	-
Acronychia pdunculata (L.) Miq.	Rutaceae
Sapindus mukorossi Gaertn.	Sapindaceae
Pterospermum acerifolium (L.) Willd.	Sterculiaceae
Sterculia villosa Roxb.	-
S. foetida L.	-
Eurya acuminate DC.	Teanstroemiaceae
Schima wallichi (DC.) Korth.	-
Microcos paniculata L. ex W & A.	Ulmaceae
Trema orientalis (L.) Bl.	Urticaceae
Villebrunea integrifolia Gud.	Urticaceae
Callicarpa tomentosa (L.) Murr.	Verbenaceae
C. macrophylla Vahl.	-
Gmelina arborea Roxb.	-
Tectona grandis L.	-
Vitex peduncularis Wallich	-
V. pubescens Vahl.	-

3.1.2. Preparation of Cultivar

Healthy, disease free and good quality betel vine cultivars are to be selected for cultivation. Farmers should have knowledge and information regarding the health and disease conditions of cultivar prior plantation. It can be collected from previously cultivated stock or other farms of more than 3 years old. The stem cutting process is used for preparing the planting stock. Generally middle aged, soft, green and fresh vines are preferable for planting. Very hard, pale yellow cuttings are discarded, because it takes a long period to emerge new shoots and sometimes fails to grow new shoot. The betel vine cutting may be of 2 or 3 or 4 nodes as required for planting (Figure 28b). Injured, twisted or broken cutting is to be rejected, since it may not grow. Usually, the planting is done on the same day of cutting preparation. If it is not possible, then the cuttings are needed to keep in shady, cold and moisture place.

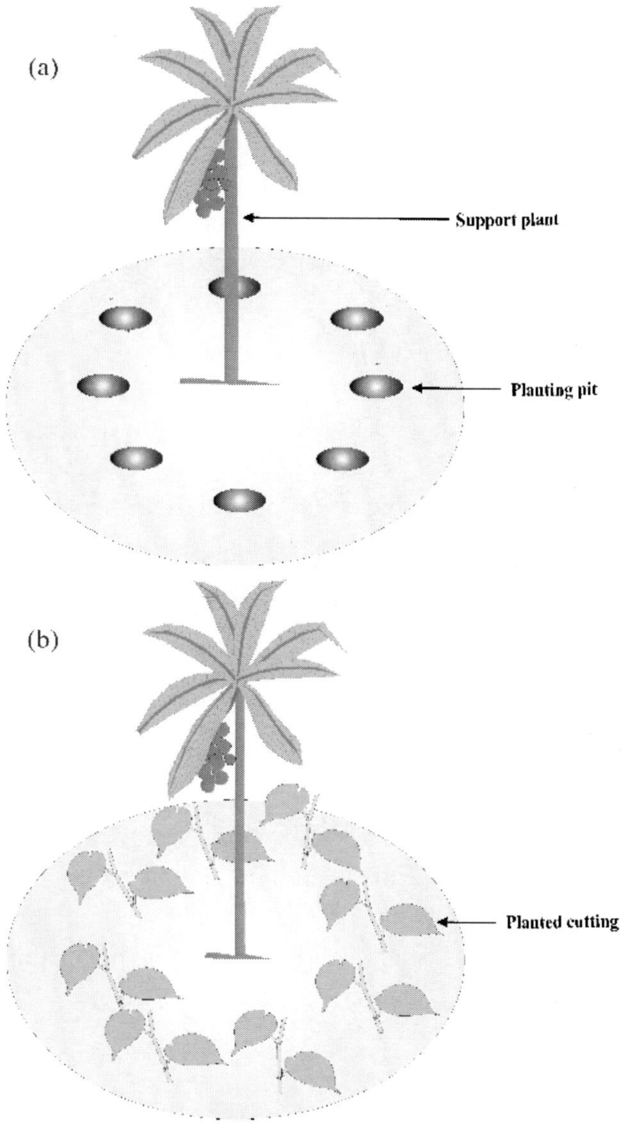

Figure 28. The procedure of betel vine plantation in open or natural environment. Preparation of pit in bed around the Areca tree (a) and planting of cutting in pit (b).

3.1.3. Planting of Cutting

Although planting season is varied with the weather condition in different places. June to August is generally suitable for planting the betel vine in hilly districts of Eastern part of India. The base of the support tree is to be cleaned by removing weeds before plantation of the cutting of the betel vine. A planting pit of about 20 cm x 20 cm x 30 cm for each cutting is required for planting (Figure 28a and b). Number of pits surrounding support tree varies according to the number cutting. For planting, two nodes of the cutting are to be kept under ground in the pit and then pit is loosely filled by soil to avoid the water stagnation in the pit.

3.1.4. Nourishing of Planted Cutting

It usually takes 15 to 20 days to emerge the new shoots from the node of the cuttings. The weeds and grass grown around the support tree and cuttings are to be removed to keep clean during this growing period. It helps to grow well the betel vine cutting. At the primary stage, the new betel vine creeps over the ground rather than to climb the support tree. It needs to tie the vine to the support tree, when it grows in length.

3.2. Management of Farm

3.2.1. Pruning of Support Tree

The pruning of support trees is to be completed within 3 to 4 weeks of planting for ensuring the sufficient sunlight and air required for new vine. Generally, about 50 to 60% of the side branches of the support tree are pruned during the period from June to August. It helps for rapid growth of betel vine and high leaf production.

3.2.2. Mulching

The debris generated from the pruning of the support tree is processed together by chopping into small pieces and decomposing at the base of the support tree (Figure 29). Weeds and grasses are needed to eradicate from the surrounding area of support trees can be used as mulch materials. These materials are gradually decomposed by a natural process. Mulching helps to retain the moister and supplies the nutrients to the new betel vine.

Figure 29. Mulching of debris generated from the pruning process at the base of support plants in order to supply as organic fertilizer for betel vine as well as support plants.

3.2.3. Manure and Fertilizer Application

Dry cow dung can be applied as organic fertilizer at the rate of 2 to 3 kg/support tree. Oil cake powder is also occasionally mixed with cow dung at the rate of 100 to 150 g/support tree. The NPK (nitrogen, phosphorus and potassium) chemical fertilizer is commonly used at the rate of 200 to 250 g/support tree for better growth. Generally, fertilizers are spread at the base of support tree and around the betel vine. Sometimes, calcium carbonate is also applied for soil refinement.

3.2.4. Irrigation

It is not common in this cultivation practice. Irrigation is usually provided during dry season (December - March) from nearby stream or natural water flows. Sometimes, shallow or deep tube well is used for watering due to avoid the problem of water scarcity. Though it is expensive, but it helps for high leaf production during the dry season.

3.2.5. Disease Management

Please vide the Chapter 9: Pest and disease control.

(a) (b)

Figure 30. Harvesting of leaf from the betel vine grown on support trees in the open or natural cultivation system. The farmer is harvesting betel leaf from upper part of support tree (a) and lower part of tree (b) using the ladder.

3.2.6. Harvesting and Processing

The peak period of betel leaf production usually starts at the age of 3 years in this type of cultivation. The life span of the betel vine is about 12 to 15 years in open cultivation practice. A special type of ladder made by bamboo pole is used for harvesting the leaf from the betel vine climbed up the support tree (Figure 30a, b). Betel leaf is usually harvested five times in a year, and the peak period of harvesting is May to January.

Figure 31. Farmers are engaged for processing the harvested betel leaves for selling in the market.

Figure 32. A natural view of a hut like structure or “boroj” or “boreja” constructed for closed or controlled betel vine cultivation farming.

Betel leaves are picked up and brought from farm to home by a special type of basket. The harvested betel leaves are arranged into small bundles and stored with, paddy straw, banana leaves, polythene sheet, cloths, etc. by spraying water to keep and look fresh (Figure 31). This arranged bundle of betel leaves are kept in the basket and sent to market for selling.

4. Closed Betel Vine Cultivation

It is a kind of widely and commonly used scientific betel vine cultivation practice under controlled conditions. It involves the cultivation of the betel vine in the plain farming land by constructing the hut like structure or closed garden, which is either square or rectangular in shape (Figure 32). This artificial structure is constructed for the development of artificial environment favourable for proper growth and protecting the vines from the adverse environmental conditions, such as, high temperature, sunlight, heavy cold, storm, heavy wind, heavy rain, etc. Though there are various names in different parts of India as well as other parts of the world, this type of constructed betel vine yard is widely called as “boroj” or “boreja” in West Bengal, India (Bhakta et al., 2016; Bhattacharya et al., 2012) (Figure 32). The different steps followed in this type of cultivation can be briefly described as below:

4.1. Establishment of Farm

Fertile soil, proper shade and irrigation are the important environmental components for successful cultivation of the betel vine. Therefore, a farm should consider these basic components for farm establishment.

4.1.1. Selection of Land

Fertile soil is best for betel vine cultivation. Furthermore, the soil qualities required for betel vine cultivation have been expensively mentioned in the subsection "habitat and ecology" of introduction. Waterlogged, saline and alkali soils as well as lack of water are unsuitable for its cultivation. Principally, the primary selection criteria of land suitable for construction of betel vine farm are fertile soil and easily accessible water resources necessary for irrigation. River, beel, lake, jheel, pond, ditch, etc. are the appropriate fresh water sources for betel vine cultivation. It requires somewhat high (about 2 ft) land from surrounding low land to avoid the water flood especially in rainy season. Besides, It should also be flat and even, well drained and with good sun shine. The field should not have a betel cultivation infected with bacterial leaf blight at least for 2 years. Various biomaterial resources, human resources (reasonable labours), easily accessible and demandable markets are also important criteria for choosing the location of land for establishing the betel vine farm. The appropriate soil criteria as well as climatic conditions as mentioned earlier are also beneficial to enhance the betel leaf production in the selected land. All these above conditions can be considered favourable in this regard.

4.1.2. Land Planning and Preparation

After selection of land, the planning is the most important step to establish a farm of the betel vine. Generally, a typical farm should have cultivated area (boroj), water resource and small farm house (Figure 33). On this basis, the planning should be done according to available land area. It also depends on the types of land and availability of the water resource i.e., whether it is low or high land and whether the water resource is available or not near land. Low plain land is needed to fill with fertile soil to avoid water stagnation. A pond should be excavated or shallow/deep tube well can be arranged for supplying water according to the requirement of farm irrigation. Generally, 2/3 part of the selected lands is allotted to construct the betel vineyard for cultivating the betel vine and remaining 1/3 part is allotted for digging up the pond or water reservoir required for water irrigation especially during winter and summer seasons. The soil obtained from pond excavation can be used for filling up the allotted plot of cultivation. It is basically a common and convenient process practiced by most of the farmers. According to size of selected land, the length, width and number of beds or rows are to be planned for better management of the betel vineyard. A small farm house is also required to keep tools and materials used for cultivation.

4.1.3. Construction of Betel Vine Yard (Boroj) and Farm House

Proper shade and fertile soil are important parameters for the successful cultivation of shade loving betel vine crop. For providing scattered sunlight, a shade is made by constructing the hut like structure called "boroj" of "boreja" (Figure 34a, b and 35a, b). A typical boroj is a shaded area of betel vine farm, which is properly fenced by the boundary. The unique features necessary within the structure of "Boroj" for proper growth of the betel vine are as follows (Kumar, 1999):

Figure 33. A typical layout plan of land to be employed for constructing the betel vine farm. It commonly comprises betel vine yard, pond (water source), vegetable garden, animal husbandry house and farm house to bring completeness of betel vine farming process.

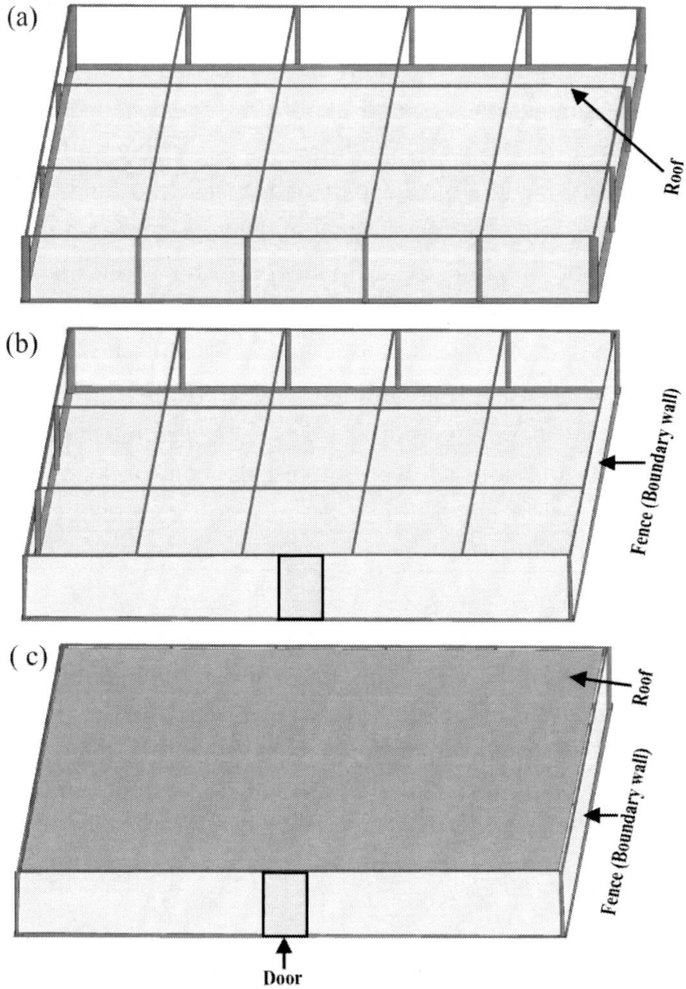

Figure 34. A standard sequential process (a, b and c) for constructing a typical hut like structure or "Boroj" or "Boreja" used for closed or controlled betel vine cultivation process. It shows stepwise construction of internal structure (a), fence (boundary wall with door) (b) and roof (c).

1. *Humidity and temperature:* The humidity and temperature are two critical factors in betel vine farming, since; the betel vine prefers high humidity and mild temperature for better growth. The structure of "boroj" should be such that it ensures for maintaining the appropriate high humidity levels and mild temperature.

2. *Incident of solar radiation:* It is one of the most important factors for suitable and high growth of betel vine plant. High sunlight damages this soft plant. Therefore, design of roof and boundary wall of "boroj" should be such that effectively inhibit the excess light and maintain the appropriate sunlight within the "boroj" necessary for favourable growth of the betel vine plant.
3. *Effective air volume:* The air volume within the "boroj" is another important factor, since it is correlated with its humidity and temperature. If the inside volume of "boroj" is large, huge amount of water is required to maintain its suitable humidity and temperature. In this respect, the appropriate volume should be maintained in designing the structure of 'boroj".

It may be either square or rectangular in shape constructed by (a) fence (boundary wall) and (b) roof and shade.

(a) *Fence (boundary wall):* The frame or structure of fence (or wall) is made by various materials such as - poles of bamboo or cylindrical wood (plant) or iron road with ropes, especially coconut rope, iron wires, paddy straw and/or jute sticks and/or coconut/date palm/other leaves and/or old clothes and/or polyethylene sheet or shade nets, etc. (Figure 34a, b, c). It is strengthened by placing the poles at specific intervals inside the boroj to protect from strong wind and storm. The four corner poles of fence are commonly stronger than other poles to give strength. Sometimes living plants (or jeol plants) are used as corner poles planted few years before boroj construction. One or more than one get(s) with movable door should be made in boroj for the convenience of work.

(b) *Roof and shade:* The frame or structure of the roof is constructed and strengthens by placing long and thin bamboo, sticks, iron wires, etc. at certain intervals on the roof and supported by bamboo poles in the inside of boroj (Figure 34a, b, c). The roof is constructed at the height of about 7 to 10 ft. from the land surface and shaded by some light-weight materials as mentioned above such as, paddy straw and/or some kinds of long grass and/or jute sticks and/or coconut/date/palm/other leaves and/or old clothes and/or shade net sheet (available in market), etc.

The shade artificially maintained the micro environment within the boroj or garden, which significantly plays an important role in maintaining the quality of leaves and in protecting the disease infection. In principle, the density of shade regulates the light intensity within the boroj or garden. The higher density shade maintains the lower light intensity inside the boroj or garden and vice versa. Besides, the requirement of light intensity for a cultivar is depended on the rate of photosynthesis, which is not same for all kinds of betel vine cultivars. The photosynthetic rate of the Meetha cultivars is lower than that of the Bangla, Sanchi, Kappori and Desawari cultivars. Therefore, meetha cultivar needs high density shade compared to that of the Bangla, Sanchi, Kappori and Desawari cultivars. It has also been reported that the Bangla variety requires high light intensity than that of Sanchi, Kappori and Desawari, because it's photosynthetic rate is higher than Sanchi, Kappori and Desawari. Thus, comparatively lower shade density is required for the cultivation of Bangla cultivar.

However, the shade should be maintained according to necessity of betel vine cultivars to be planned for cultivation. Since, no such systematic research work has been performed to find the cultivar specific density of shade so far. An extensive research work is also required in this respect for qualitative and quantitative improvement of the betel vine. Recently, a kind of shade net has been developed and available in the market, but shading percentage of the net should be standardized for cultivars before application, otherwise it may cause adverse impact on the betel vine.

Generally, farmers use the locally available materials for constructing a boroj (Figure 35a, b). The types of materials used in constructing the boroj are widely varied in different parts of India as well as other countries of the world. The following materials are commonly required to construct a boroj of 5 decimel area of about 7 to 8 feet height that may cost roughly around twenty thousand rupees (about US $ 400):

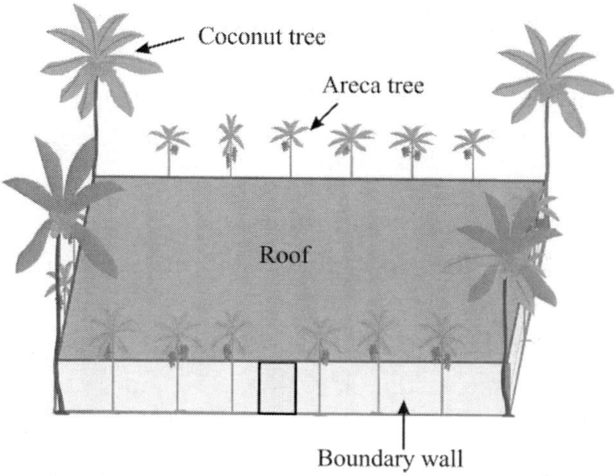

(a)

(b)

Figure 35. (a) A pictorial representation of the typical hut like structure or "boroj" or "boreja" and (b) Natural views of a constructed "boroj" or "boreja" for closed or controlled betel vine cultivation system. The coconut trees are planted at the four corner of the"boroj" or "boreja" and Areca trees are planted at the side of four boundary walls for gaining high benefit in addition to the betel vine cash crop.

1. Bamboo : 16 pieces
2. Iron wire : 28 - 30 kg
3. Coconut rope : 8 - 10 kg
4. Paddy straw : 480 bundles
5. Jute stick : 10000 pieces
6. Bamboo pole : 1000 pieces

A farm house can be prepared using the bamboo poles, coconut ropes, paddy straw or plant's leaves and other plant materials as a store room for keeping the materials and instrument used in cultivation. It can be used as a rest house of workers during working period in the daytime.

4.1.4. Soil Preparation and Sterilization

The types and soil qualities required for betel vine cultivation have been discussed in the subsection "habitat and ecology" of Introduction. The preparation of betel vine bed with fertile and sterilized soil is the most important aspect for successful plantation. The top soil is dug and oil cakes, cow or poultry dung, rotten farmyard manure (FYM) (Thomas et al., 2013) and leaves are thoroughly mixed or incorporated well in the topsoil by tilting with the help of plough or power tailor. The newly dug up clay soil should be allowed for sunlight for about one year to sterilize before utilization in the betel vine cultivation. The soil should be tilted several times (about 4 – 5 times) and allowed for sunlight for few days before providing the roofs shade. This process is called as solarization, principally for soil sterilization. The process of solar sterilization includes the following steps (1) a transparent and thin polythene sheet is covered over the soil for about 30 days period during the summer season to increase the temperature in order to kill the insects, pests and pathogens and (2) Spray water on soil once per week towards afternoon removing the polythene and again quickly cover it just after spraying water. This process reduces the adverse effect of pathogens and helps to establish the healthy betel vine.

Although, it is not recommended for causing environmental pollution, in case of new farm, the soil can be sterilized by using some insecticides, pesticides and antipathogenic agents. For example (1) carbofuran 1.5 kg/ha or 0.75 kg/ha mixing with the 500 kg mastered oil cake can be recommended for controlling nematodes and (2) The solution of formalin and water (1:50) is sometimes applied in soil, and a transparent polythene is covered over soil for about 5 to 6 days period. After about 25 to 26 days of opening the polythene sheet, the cutting of the betel vine should be planted. In case of old betel vine farm, the application of carbofuran should not be recommended, because it is one of the most toxic carbamate pesticides. Therefore, the betel leaves not be picked up before 65 to 70 days of carbofuran application.

4.1.5. Preparation of Betel Vine Bed

The types of bed for betel vine cultivation varied largely from place to place (Figure 36 a, b, c). Keeping a 2 to 3 ft working path alone the inside of the four boundary walls, the bed should be prepared. In intensive cultivation, about 30 to 45 cm width x 20 cm height bed is prepared with prepared and sterilized top soil at about 0.6 ft interval by making furrow of about 20 to 45 cm width x 20 cm depth between two beds. The length of bed is according to the length of the boroj. Besides, the size of the bed is varied in different places with the space available, for example, 1.0 to 1.25 m wide and as long as the entire length of beds is prepared, leaving about 0.3 m walking path between two adjoining beds. Sometimes 1.2 m × 7.5 m beds are prepared. Adequate spacing should be left between beds to allow management practices and to control the disease infection. Beds should be sterilized by burning straw or ash on it. Around the cluster of beds a drainage canal of 30 cm width, 60 cm depth should be built.

An artificial live or dead support should be provided to betel for upright climbing. Supports, called as stakes, are established or placed in the beds at the spacing of 0.45 m × 0.45 m.

In some cases, integrated faming of betel vine and areca nut is found. In this case, betel vine is cultivated as a mixed crop in the drainage channels of areca nut cultivation fields. The drainage channels are made between every two, three or four rows of areca nut depending upon soil type. The trenches having depth 30 to 60 cm are prepared 1.0 to 1.5 m away from the base of supporting tree. In some places of India, Shevri or Rangara is used as supporting tree. It is tilled with top soil plus 10 kg FYM (farm yard manure) spacing is given according to the need of supporting tree.

There are different methods of cultivation depending upon soil, climate, irrigation facilities, etc. (1) Basin, (2) Flat bed, (3) Dig trench, (4) Long bed trench (5) Short bed trench, (6) Trench and furrow system.

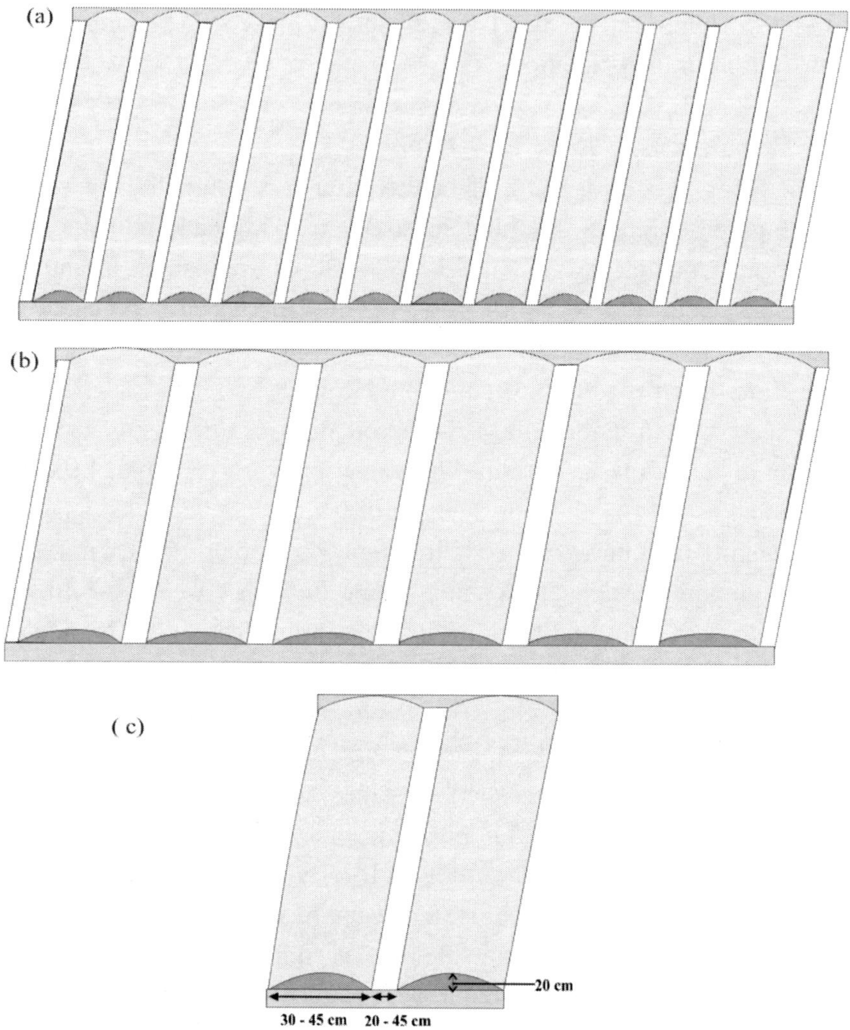

Figure 36. Figures depicting the beds of the betel vine in the closed or controlled cultivation system. It illustrates the beds of intensive cultivation (a), normal cultivation (b) and shape and size of bed (c).

The size of the bed is 3 × 4 m. There is the provision of irrigation channel and sub-channel with-suitable, drains. After layout in the middle of June, seeds of support plant, Shavri are sown in order to support the vine along the side of beds. Pangare *(Erythesna indicd)* is also planted at the spacing 1.5 to 2.0 m along the border of garden.

4.2. Crop Establishment

Selection of Cultivars

For beneficial and sustainable cultivation, the betel vine cultivars having the following characteristic features are to be chosen:

1. Highly tolerance to adverse environmental conditions
2. Comparatively high market value to get good price
3. Comparatively high demand in the market
4. Fast growth properties
5. High yielding capacity
6. High disease resistant
7. Leaves of comparatively bigger and with good taste quality
8. Low fibrous leaf
9. High leaf preservation efficiency
10. High essential oil content

Table 12. The seasons for planting betel vine in different states of India

States	Seasons
Andhra Pradesh	September – October
Assam	April – May, August – September
Bihar	June – July, August – September
Karnataka	July – August
Kerala	May – June, September – November
Madhya Pradesh	January – March, September – November
Meghalaya	August – September
Nagaland	April – May
Orissa	February – March, May – June, September – November
Tamil Nadu	March – May, August – October
Tripura	March – April, September – October
Uttar Pradesh	October – November
West Bengal	February v March, June – July, September – October

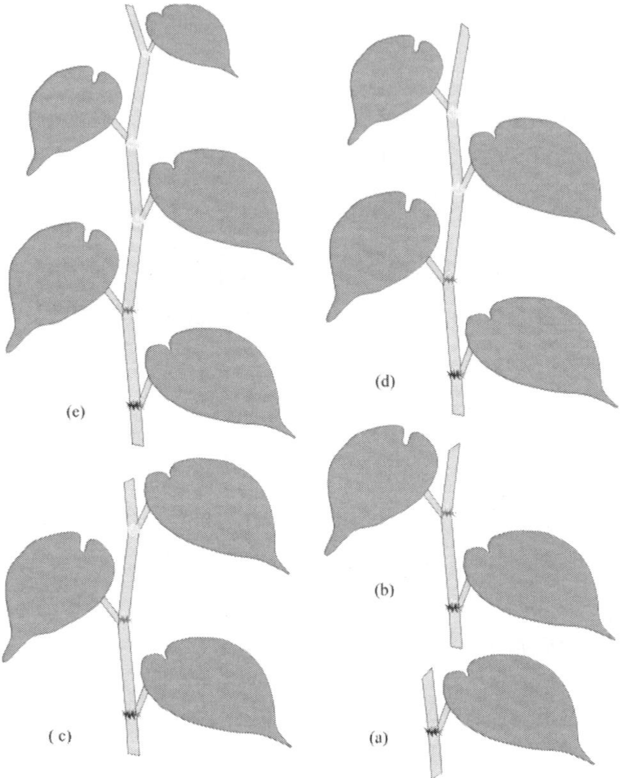

Figure 37. Preparation of cutting of 1 node (a), 2 nodes (b), 3 nodes (c), 4 nodes (d) and 5 nodes (e) employed in the closed or controlled cultivation system.

Season of Cultivation

The planting season is varied in different countries and in even different parts of same countries. The humid environment is favourable for high growth of the betel vine. The betel vine is generally planted in rainy season. It is months of June and July, at the beginning of the monsoon season in West Bengal, India whereas in north India the planting season is October to Nivember and in south India it is May to August. The seasons of betel vine plantation in different states of India are shown in Table 12.

Cutting Preparation

The propagation of the betel vine is obligatory only through vegetative means. It is usually propagated by using stem cuttings. Healthy and disease-free mother vines of about 2 to 5 years old are collected from authentic sources and used for cuttings. The betel vine should be treated by spraying any fungal disinfectant, such as 0.4% bordeoux mixture before using as cutting. The betel vine stem is cut just at below of the node keeping 1 to 4 node(s) per cutting (Figure 37a, b, c, d). Sometimes entire vine having 10 to 15 nodes is planted without cutting. This process is specifically followed in case of Meetha cultivars in West Bengal, India. The short joined cutting in which leaf joints or nodes are at a short distance obtained from sufficiently matured plants are always better than the cutting in which the distance between the leaf joints is more and from lower portions of the vines. If the cutting is too long, it should be shortened from the tip and not the base. Normally, a cutting should be planted immediately after its removal from the mother plant. The lower nodes of the vines take longer time to sprout or even fail to do so due to deep dormancy of the adventitious roots.

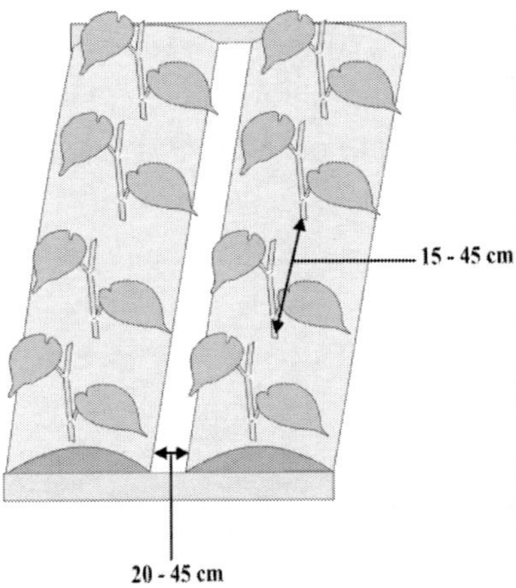

(a)

(b)

Figure 38. Plantation of betel vine cutting in the prepared bed. It depicts the space between the cutting planted in bed of standard shape and size (a) and natural planted cutting in bed (b) used for closed or controlled cultivation practice.

Sterilization of Cutting

As the betel vine is vegetatively propogated, special care should be taken for disease-free planting material to check the sett borne diseases. The creeper cuttings are planted after proper dressing of bed and cutting. Before planting, cuttings should be immersed for about 15 to 20 minutes in a fungicide mixture (1% bordeoux mixture or 0.1% ceresin and streptocyline 500 ppm).

Planting

The cutting is planted at about 2 to 3 inches below the soil surface and nodes are filled with top soil in bed (Figure 38a, b). In intensive cultivation, the gap/space between two cuttings is varied from 15 to 45 cm in each row having 45 to 60 cm distance between two rows having 20 to 45 cm walking path. The spacing for cutting varies from the region to region and types of cultivation. In north India, it is 12 to 15 cm × 75 to 80 cm, but in south India, it is 100 × 20 cm spacing. In Sundarbans of West Bengal, India, the planting spacing is generally 60 × 30 cm. In some other types of cultivation, cuttings are planted in 30 × 30 cm pits, filled with top soil and cow dung mixture, and stakes of 2 to 4 cm diameter should be established as supports. Either live supports of *Gliricidia sepium* or durable dead wood support can be used. Two cuttings are planted near a stake. Setts with vigorous apical buds and nodal adventitious roots are selected and planted at the base of the live supports, which are to be planted 4 to 5 months earlier. The cuttings are planted in the furrows (8 - 10 cm deep) of sterilized soil with spacing of 50 to 60 cm × 10 to 20 cm during rainy or autumn season. Cuttings can directly be planted in field or can be planted as rooted cuttings established in poly bags filled with a mixture of equal parts of top soil, cow dung, coir dust and sand. In some places, the entire betel vine of about 5 to 7 ft length is directly planted in a row. This type of practice is generally followed in Meetha cultivars planted in West Bengal, India. In that case, the number of plants required for each ha is lower compared to that of cutting process. Such, on an average 10,000 plants are required for planting in one hectare area. The number of cutting required for planting in I ha of farm varied with spacing between two rows and number of vines to be planted, which is shown as follows (Table 13):

Table 13. The number of cutting required for planting in one ha farm area under different spacing

Row spacing	Vines/hectare	
	Single vine	Double vine
20 cm	50,000	1,00,000
30 cm (1 ft)	30,000	60,000
45 cm (11/2 ft)	22,500	45,000

4.3. Management of Farm

Nourishment

Immediately after plantation, the specific careful nourishment process should follow, which is the important factors for ensuring hundred percent survival of planted cutting. The cutting is immediately covered with the wetted paddy straw or coconut fronds or other shading material for about 4-6 weeks to protect from direct sunlight. Beds are watered once or twice everyday using sprayer depending on weather condition. Sprouting of shoot from cuttings starts within 20 to 45 days and shade should be removed gradually afterwards. Either live supports (Such as, *Gliricidia sepium*) or durable dead wood support can be used to climb the new betel vine creeper. Eradicate the weeds and grass from the bed carefully without disturbing the planted cutting.

Other Management

The management of sophisticated betel vine farming is simple and laborious, but very much important. It needs intensive nourishment and takes care throughout the year in order to obtain the high production benefit. A little negligence in management leads to serious damage of farm ultimately resulting in the low production as well as loss farm benefit. Therefore, management is a crucial aspect in betel vine farming practice which has elaborately been discussed in the "Farm Management Technology" (Vide Chapter 8).

REFERENCES

Alam, M. K. and Mahiuddin, M. (1995). Conservation of the diversity through betel-leaf (*Piper betle* L.) based agroforestry in Sylhet. *Bangladesh Journal of Forest Science,* 24(2): 49-53.

Bhakta, B., Bhakta, U. Ghanti, S., Jana, B. B., Ghanti, P., Yamane, S., Bhakta, J. N. (2016) Betelvine: A green gold for sustainable economic upliftment in rural India. *Int. J. Env. & Tech. Sc.* 3:76-82.

Bhattacharya, R. Mondal, B., Ray, S. K. and Khatua, D. C. (2012). A study on bacterial disease of betelvine in West Bengal, India. *International Journal of Bio-resource and Stress Management,* 3(2): 211-216.

Costa, T., Datta, A (2007). The Khasia of Bangladesh; A socioeconomic survey of Khasia people. Society for Environment and Human Development (SHED), Dhaka, Bangladesh, 59.

Fila, P., Haider, M. R., Mohiuddin, M., Zashimuddin, M. and Alam, M. K. (2006). Kashia betel-leaf cultivation technique (in Bangla). Bangladesh Forest Research Institute, Chittagong, Bangladesh, 12.

Haider, M. R., Khair, A., Rahaman, M. M. and Alam, M. K. (2013). Indigenous management practices of betel-leaf (*Piper betle* L.) cultivation by the Khasia community in Bangladesh. *Indian Journal of Traditional Knowledge* 12: 231-239.

Kumar, N. (1999). Betelvine (*Piper betle* L.) cultivation: a unique case of plant establishment under anthropogenically regulated miclimatic conditions. *Indian Journal of History,* 34(1): 231-239.

Nath, T. K., Mokoto, I., Islam, M. J. and Kabir, M. A. (2003). The Khasi tribe of North Eastern Bangladesh: Their socioeconomic status, hill farming practices and impact on forest conservation. *Forest tree and livelihoods* 13: 279-311.

Thomas, U. C., Chandini, S. and Thomas. A. (2013). Integrated Nutrient Management Studies in betelvine *(Piper betel L.). International Journal of Applied Research and Studies*, 2(7).

Chapter 8

FARM MANAGEMENT TECHNOLOGY

1. INTRODUCTION

The betel vine is a quit sophisticated plant rather than other cultivated plants. Therefore, cultivation technology of the betel vine is highly specialized and needs intensive care and management. Its growth is largely influenced and hampered by fluctuating temperature, water irrigation and sunlight. Considerable efforts of management for newly grown betel vine sapling are required to adopt the new environmental and climatic conditions, since betel vine is sensitive to soil and climatic conditions. It needs proper maintenance and management in different aspects for its growth, such as, sunlight, irrigation, moisture, sufficient humidity, soil fertility and roof and side fencing. The betel vine is grown in traditional farming systems many of which are managed exclusively or communally. Therefore, the crop demands regular and continuous care and monitoring.

2. Management Practice

The following management practices are generally required in beneficial and sustainable cultivation of the betel vine:

- Sapling nourishment
- Irrigation
- Manuring and fertilization
- Training and pruning
- Lowering and earthing up
- Weed control
- Leaf quality management
- Harvesting, post-harvest, packaging and marketing
- Crop rotation

2.1. Sapling Nourishment

The cutting or sets are established within 3 weeks and the first leaf comes within a month. The new sprout starts rapid growth in the congenial environmental conditions during this period. The weeds and grasses grown on the bed or around the support materials or trees and cuttings are to be eradicated to keep clean the bed. It also helps to maintain the nutrient condition of the soil and to grow well the betel vine. At the primary stage, the new betel vine creeps over the ground rather than to climb up the support material or tree. An artificial live or durable dead support (jute stick, split bamboo stick, etc.) should be provided to each betel sapling for upright climbing by twining around upright sticks (Figure 39). It needs to tie the vine with support at about 4 to 6 weeks for proper growth and to protect the stem and leaves from damage caused by fall down.

(a) b)

Figure 39. The new betel vine is tying with support material (a) and soiling (b).

2.2. Irrigation

The betel is a vine of very soft nature and sensitive to surrounding environmental conditions, especially water content of soil and intensity of sunlight. It grows in the top soil of land, which is rapidly dried by water evaporation. Therefore, regular monitoring the water content in top soil is important. The plant's growth may hamper, or they may die within few days for lacking of water in soil. Betel needs constantly moist soil, but there should not be excessive moisture. Hence, irrigation is a common practice in betel vine cultivation.

The frequency of irrigation also largely depends on some factors such as - season, types of soil and place of cultivation. Except rainy season, irrigation is essential in winter and summer months. In dry summer months, irrigation is given frequently. During winter months, the frequency may be at 3 to 5 days interval and in rainy season at 8 to 10 days interval based on the rain and soil condition. The irrigation interval is long in soil type having high water-retention capacity and vice versa. Quantity of irrigation should be such that the water should not stand over the bed and even if water standing in the walking path should not remain for more than half or one hour. If water is logged by heavy rains or excess irrigation, the water should be immediately removed through drainage system or by the water pump. The best time for irrigation is morning or evening. The process of irrigation should be gentle as the leaf is not damaged by soil/clay as well as a pathogen to avoid disease prevalence. A good arrangement for drainage of excess water is always essential to save the roots from decaying and drooping of leaves (Chandra and Sagar, 2004). The water of neutral pH (about 7.4) is suitable for betel vine irrigation. The application of wastewater and salty water for irrigation should be avoided, which may hamper the growth of betel vine.

There are several irrigation devices, viz. pitchers, pot of tin, water pumping machine, etc. Previously, the irrigation process was troublesome matter, because not only the betel vineyard and also vegetable and flower garden is irrigated only by hands using the earthen pitchers or pots of tin i.e., hand operated irrigation. The water-filled pitcher or pot is carried by hands from the water source (ponds/other water body) using chin wise man power system during the long period. Now, the betel vineyard is easily irrigated by water pumping machine from the pond or deep/shallow tube well within a short period. The water pump may be of photovoltaic/ diesel/electric power driven. The irrigation using a water pump is a costly process compared to that of the hand-operated irrigation.

2.3. Manuring and Fertilization

The betel vine is a perennial cash crop cultivated for its green leaf production year-round. For that why, the betel vine cultivar with fast growth property is required to obtain the high yield of leaf throughout the year. Besides, betel vine is very fastidious regarding its manorial requirement. It efficiently absorbs nutrients from the top soils by its adventitious roots and maintains growth. Therefore, it is apparent that growth of betel vine creeper is principally influenced by the nutrient's condition of the top soils used in cultivation along with other environmental conditions. The nutrient lacking may be occurred in the soils after few months of cultivation due to its continuous nutrients uptake from top soil. In this respect, it should be remembered that the dose of fertilizer should be based on nutrient contents of soil, which can be determined by analyzing the nutrient's quality of soil. Such, the problems of lacking or excess nutrients can be avoided. Applications of manures and fertilizers at a certain interval are the basic strategies to maintain the qualitative and quantitative fertilities of the soil in order to achieve high growth of the vine as well as high yield of leaf. In spite of these, qualitative nutrient's content in the manure or fertilizer, i.e., the ratio of nitrogen (N), phosphorus (P) and potassium (K), play a pivotal role in the growth performance of the betel vine. The nitrogen plays a direct role on growth, yield and keeping quality of betel leaf. Moreover, nitrogen availability throughout the growth period of crops from the applied manures is also of considerable importance.

The betel vine is a highly labour intensive crop and a voracious feeder of nutrients requiring about 400 to 600 Kg N/ha, 200 to 300 Kg P2O5/ha and 200 to 250 Kg K2O/ha per annum, which are commonly supplied only through the organic sources. In some places, inorganic (chemical) fertilizers are also being used sparingly now-a-days. One hectare of the betel vine yard required NPK at the ratio of 160:80:40 Kg/ha (Chandra and Sagar, 2004; Thomas et al., 2010; 2013). The organic, inorganic and mixed of both organic and inorganic fertilizers can be applied in the cultivation of betel crop (Thomas et al., 2010; 2013; Mandal et al., 1994) to proper maintenance of nutrient supply throughout the year.

The organic fertilizers, such as - farm yard manure (FYM), cow dung, goat manure, poultry manure, oil cakes (neem, mustard, groundnut, coconut, etc.), fish mill, rice bran, vermicompost, etc. are mainly apply in this crop (Chandra and Sagar, 2004; Thomas et al., 2010; 2013). The farmers usually apply heavy doses of bulky organic manures in the betel cultivation. The manures and cakes should be well composed and dried before application to plant; otherwise, it may cause adverse effect. Principally, manuring should be started with top dressing with oil cake. Initially, cow dung or compost should be spread over the bed at the rate of 50 to 60 ton/ha/year after about one month of the newly planted cuttings. The manuring after every harvest of crop is a very helpful practice to maintain the proper health and continuous production of leaf. Betel crop needs heavy manuring of farm yard manure (FYM) usually at the dose varied from 35 to *50* ton/ha/year along with oil cake, fish, etc. Using 250 gm of neem cake/plant is highly advantageous since it will also reduce the pest incidence. Application of decomposed glyricidia leaves is highly beneficial for better growth and higher yield. One month after the planting of cutting or setts, oil cake should be applied at 15 or 30 days interval, along the rows and about 8 to 10 cm away from the vines and mixed in the soil. During summer, apart from the oil cake, some seeds and leaves of mango, karanj and neem can be used. Dried neem leaf or C*alotropis* leaves at 2 t/ha and cover it with mud (2 t in 2 split doses) can also be applied for good health of plant (Chandra and Sagar, 2004; Thomas et al., 2010; 2013).

The inorganic fertilizers, such as – urea, diammonium phosphate (DAP), super phosphate, potassium sulphate, etc. are widely recommended for cultivation of the betel vine (Table 14). The inorganic fertilizers can be applied in bed at about 15 to 20 cm away from the betel vine. The inorganic fertilizer is applied as basal dressing one month after the planting at the following rates:

The application of NPK fertilizers in the proportion of 50 : 25 : 25 g/tree/year immediately after lowering (Thomas et al., 2010; 2013). Betel leaves are picked up once in every 3 to 4 weeks, and with that substantial quantity of nutrient is removed from the soil. Therefore, application of chemical fertilizer is essential for higher yield and better growth. The recommended mixtures of chemical fertilizer are as follows:

Table 14. Inorganic NPK (Nitrogen, N; phosphorus, P; and Potassium, K) fertilizer schedule for betel vine cultivation

Time of application	N	P	K
	(kg/ha)		
Basal dressing	37.5	100	50
Top dressing @ 3 split doses	112.5	0	0

Urea	195 g
Triple Super phosphate	65 g
Muriate of Potash	100 g
Keserite	60 g

The above mixture should be applied at the dose of 420 g to 100 betel vines in every three-week intervals.

2.4. Integrated Nutrient Management (INM)

Farmers do not like to use commercial chemical fertilizers, since they think that the quality of the leaves may be affected by using the commercial fertilizers. Adverse effects of chemical fertilizers on keeping the quality of leaves and aggravating foliage diseases have also been reported (Das et al., 1989; Mandal et al., 1994). In this point of view, the term integrated nutrient management (INM) has evolved recently. It refers to the integration of organic and inorganic fertilizers, which is invariably advantageous from a long-term perspective both in terms of cost of production as well as soil health in betel vine cultivation. However, considering the cost of production and soil health, INM is always advantageous for long-term benefits (Debanath et al., 1985). In this respect, it is pertinent to mention herein that the integration of inorganics and organics fertilizer is responsible for highest yield, and this supports the concept of integrated plant nutrient management approach (Thomas et al., 2013). A study conducted at College of Agriculture, Vellayani, showed that the bulky organic manures can be partly substituted by chemical fertilizers without affecting the chewing quality of leaves (Chandini, 1989). From this study, it was also recommended to manure the betel vine crop with 60 kg N ha^{-1} and 25 kg P_2O_5 ha^{-1} over a basal dose of 30 ton ha^{-1} of farm yard manure (FYM). Maiti et al. (1995) reported that application of 200 kg N ha^{-1} through organic and inorganic sources in 1:1 ratio is the best way for obtaining higher yield in betel vine.

Additionally, some scientists also recommended the following fertilization schedule: total 150 kg N/ha/year through Neem cake (75 kg N) and Urea (75 kg N) and 100 kg P_2O_5 through single super phosphate (SSP) and 30 kg Muriate of potash (MOP) in three split doses, firstly, at 15 days after lifting the vines and second and third doses at 40 - 45 days intervals.

2.5. Use of Plant Growth Regulators

The different research experiments have observed that betel vine growth and yield can be increased through the application of plant growth regulators. Three sprays of Mixtalol (0.1%) and Tricontanol (0.5%) were found very effective in increasing the growth and yield of the betel vine. It is also capable to alleviate water stress damage under medium stress. In this respect, more studies are required.

2.6. Training and Pruning

After about a month of planting, the cutting began to sprout and creep. The betel vine is generally trained either to live supports in open cultivation or dead supports in closed cultivation, but concrete post or coir ropes can also be used as substitutes. In intensive cultivation, a thin support material such as bamboo or jute sticks is placed at the side of the betel vine to climb up. The vines trail up to 1.80 to 2.10 m by 6 to 8 months period. They are fastened by a kind of grass (Figure 40) with the supports usually at every 20 to 30 days interval to avoid damage may cause by fall down of vine. This operation has to be performed at least 5 to 6 times for Bangla, Meetha and Sanchi varieties during every subsequent year, especially in the closed betel vine cultivation. The growths of the vines are rapid during the rainy season and slow during winter season (Chandra and Sagar, 2004).

Figure 40. Farmer engaged for training the betel vine.

The pruning process is preferred at 1 m height of the betel vine to increase the plageotropic branches and yield. After 1.2 m growth of the betel vine trellis is established. It is generally followed in the tree-based open betel vine cultivation for increasing the side branches in order to enhance the production of leaf.

2.7. Lowering and Earthing Up

Under normal cultivation, the vines grow to height of 3 m in the one-year period. When they reach this height their vigour to produce normal size leaf is reduced, and they need rejuvenation by lowering. The lowering is a very common management practice, especially in case of intensive closed betel vine farming (Figure 41). It is typically a cyclical process usually followed 2 times per year. Lowering of vines is done in quick succession during the rainy season. When the vines reach to (a height of about 1.80 – 2.10 m) roof shade, they not only loose their vitality but also become difficult to harvest the leaves from upper part of the vine. There is a high chance to damage the leaves at growing end in this stage. The vines are often regulated by giving fold or coiling, which is known as "Lowering" (Figure 41). Keeping 4 to 6 young leaves (i.e., 4 – 6 nodes) at the growing end of the vine, remaining leaves and small branches are picked up before lowering. It is a process in which vines are released/relaxed from their support, brought down, coiled/folded the lower portion (leafless) on bed and again tied the upper growing end of the vine with the lower part of the corresponding supports. Consequently, vines get the opportunity to creep over the free space of supports. After the vine is lowered, the more tillers spring up from the nodes of the stems coiled/folded on the bed and produce many primary vines. This helps in vigorous growth of vines. The damaged support materials are replaced and rearranged by new one during the process of lowering. Branches developed from the mother vine are also tied with new support materials resulting in increment of the betel vine's number in each lowering process. To get the quality and large-sized leaves, the number of vine should be standard and limited per unit space. It may be 15 to 20 in number per meter row of bed. For that why, small, unhealthy and old vines producing poor quality leaves should be removed from bed during lowering process. The beds and walking lanes are dirty by unwanted leaves, small branches, old vines, damaged support materials, etc. by this entire process of lowering and weeds grown on bed and lane. The process of cleaning the bed by removing garbage and eradicating the weeds can be termed as "dressing".

After finishing the process of lowering and dressing, the coiled or folded stems of the vine are buried in the soil or covered by soil, which is called as "earthing up". Soils are usually stored near the boroj prior using in earthing process. In case of emergency, paddy straw is used for covering instead of soil for short period to overcome the problems of man power shortage or shortage time or lack of soil, etc. Sometimes, especially in the winter season, the coiled or folded stem is kept in naked condition without covering the soil or paddy straw to develop the new healthy shoots from the coiled stem in order to increase the number of vines as well as production of leaves. After sprouting shoot, the coiled naked stem is covered by soil. However, height of the beds goes on increasing due to frequent earthing of the vines. The vines are grown on elevated beds imitating the natural ecological conditions suitable to the crop. Besides, this elevated bed is helpful to avoid the problem of flood disaster. Irrigation should be given after each lowering. Manuring is done after each lowering and earthing up process.

(a)

(b)

(c)

Figure 41. Lowering process of the betel vine: (a) opening the vine from the support stick, (b and c) binding the vine with a stick after folding on bed.

Figure 42. Earthing up or soiling or covering the betel vine stem by soil after lowering.

2.8. Weed Control

A weed is a plant considered undesirable in a particular situation which competes with productive crops or pasture for space, nutrients, water and light, ultimately converting productive land into unusable scrub and leads to ecological and economic impacts. Weeds can be poisonous, distasteful, produce burrs, thorns or otherwise interfere with the use and management of desirable plants by contaminating harvests or interfering with livestock. Varieties of weeds such as - grass, unwanted creepers can grow in the betel vineyard which may compete with betel vine when betel vines grow in the bed and climb up on the support tree or materials employed for betel vine. The soil should be treated and sterilized properly to protect the problems of weeds before establishment of the vineyard. Besides, weeds may grow rapidly on the shade of the roof, which alters the light intensity within the boroj. After planting, it usually takes 3 to 4 weeks to emerge the new shoots from the node of the cuttings. The grass weeds grow around the support tree or dead support materials and cuttings during this period, which may also cause negative impact on the growth of new sapling. Proper *weed management and control* are required for proper growth and high production of leaf. It is not a critical challenge for the betel vine cultivation. The farmers need to monitor regularly, removed weeds and always keep clean of the bed and walking path within and surrounding the boroj. It helps to grow well the betel vine sapling. No recommendation can be made for applying the herbicides for controlling the weeds within the boroj, because, it easily damages the betel vine along with weed.

2.9. Leaf Quality Management

Quality maintenance of marketable leaves is the importance aspect of betel vine cultivation. Since, the market value of leaf is largely varied with leaf quality and hence obviously affects the cost-benefit of farm. Sugar, essential oil and other active compound's content are important chemical parameters determining leaf quality in the betel vine. Good taste is also another property of the good quality leaves for chewing purpose. There are no specific quality parameters for betel leaves, the good export quality leaves are supposed to be characterized by the following external properties which play important roles for marketing the leaves:

1. Size of leaf – large size (at least 20 cm in length and 15 cm in width)
2. Shape of leaf – heart shaped (no deformative leaves)
3. Colour – attractive well matured dark green or greenish-yellow
4. Freshness and brightness – very fresh and bright
5. Pungency – low pungency
6. Thickness – thicker
7. Life span – long

The leaf quality is essentially maintained by the influence of soil quality and types of fertilizer applied along with other environmental conditions of the cultivation process. Organic nutrient sources significantly impact on good quality of betel vine leaves. Mandal et al. (1994) reported that complete reliance on chemical fertilizers without organic input could produce inferior quality of leaves and adverse effects through an aggravation of foliage diseases. Supplying N completely through chemical fertilizers generally led to a more rapid release rate for N, producing more succulent leaves and therefore, a shorter shelf life. The substitution of inorganic sources with organic fertilizers helped to reduce the N supply rate and provide more appropriate quantities throughout the crop's growth stages. The study showed INM has significant influence on essential oil content of leaves. Thomson et al. (2010) proposed that the essential oil content is highest in the betel vine cultivated by applying 90 kg N/ha. Rahman et al. (1990) reported an increase in essential oil content in coriander seeds with increasing N application from 0 to 60 kg/ha (Thomson et al., 2010).

2.10. Harvesting, Post Harvest, Packaging and Marketing

2.10.1. Harvesting

Initial harvesting of leaves starts from about 3 months period after plantation, when vines grow about >8 to 10 leaves. Once harvesting starts it continues until they grow and produce leaves throughout life period of about 12 to 15 years. Keeping about 4 to 6 leaves at the upper growing end of the vines, comparatively mature leaves are generally picked up from the lower part of the main stem and thereafter from both main and lateral stems. Period of harvesting primarily depends on the growths of vine, which in turn greatly influenced by the climatic conditions, soil quality, fertilizer application and varieties of the vine. The production is higher during the rainy season. Moreover, interval of harvesting depends on the demand of market. Though there is no specific time interval, the leaves can be picked up from vines once or several times between two lowering periods of the betel vine. Harvesting can be done at 15 or 30 or 60 days intervals depending on the high market demand to get high benefit. The continuous picking up process promotes to vines to produce a higher number of leaves if nutrients of soil are sufficient for growth. The production lasts for several years from the date of planting. Generally, crop yield is less in first year, maximum in middle and less towards the end of life span of vines.

The betel leaves are picked from the vine by the nails of thumb finger or artificial nails, since petiole of leaf is easy to break. The thumb mail or artificial nail is usually pressed on petiole at the place just between other fingers of the hand to pick the leaves from the vine. It can be done by both hands to pick a large number of leaves at a particular period. The leaves are kept in the special type of basket and carried to farm house or from farm to home by a special type of basket. The leaves are arranged into small bundles usually containing 40 to 50 leaves and stored with wrapping materials (i.e., banana leaves, polythene sheet, etc.) by spraying water to keep and look fresh in the same types of the basket (Figure 43). Moreover, the picked leaves are washed, cleaned, counted and sorted into different grades according to size, colour, texture and maturity and their chewing quality.

(a)

(b)

(c)

Figure 43. (a) Farmer is arranging the picked up betel leaf, (b) arranged betel leaf and (c) arranged leafs in bamboo basket are ready to bring market.

2.10.2. Post Harvest

The Betel leaf is a very perishable commodity and therefore, always subject to wastage by quick spoilage due to dehydration, fungal infection, dechlorophyllation, etc. The post harvest care should take to maintain leave quality as well as to avoid huge damage caused by spoilage of leaves. The damage may range from 35% to 70% during transport and storage. In India, even if the most conservative estimates of 10% loss were considered that too would reveal a menacing loss of betel leaves worth about Rupees 900 million every year. Not only this, particularly in the rainy season a large portion of the leaves remains unsold or sold at a throw away price. In view of the alarming losses, attempts are being made to minimize the wastage by drying the leaves, controlling senescence by chemical treatments, manipulation of storage temperature, adopting better packaging materials and methods besides curing and bleaching of the leaves. Besides, wastage may also be minimized by extracting essential oil from the inferior or unsold leaves.

2.10.3. Packaging and Marketing

The arranged leaves bundles are together arranged within a basket generally made from bamboo and tightly packaged with paddy straw, cloth/jute sheet and rope. This packaged is ready to transport for local as well as distance markets.

Marketing of betel vine crop is quite complex and risky due to its perishable nature as above mentioned, seasonal production and bulkiness. The spectrum of prices from producer to consumer, which is an outcome of demand and supply of transactions between various intermediaries at different levels in the marketing system, is also unique. The marketing strategies of the betel vine are chain wise system from farmers to consumer (Figure 44). The carefully arranged betel leave's bundles are sent to local market for sale. There are many intermediaries worked as a buyer in the local market for purchasing the leaves form farmers. These intermediaries again sent these packed leaves to different companies engaged for marketing in national and international levels. For the international exporting market, the bundles are packed in specially prepared cane baskets. The leaves are exported to the USA, different European countries and some other countries where betel vines are not cultivated.

In marketing, the important problems are (1) intermediaries buy the leaf product with low and not reasonable values to obtain their much benefit, (2) lack of proper and established market with reasonable and standard value for selling the betel leaf, (3) lack of proper transport/communication from product place to market, (4) lack of cold storage for preservation to over come the unfavourable market and environmental conditions and (5) lack of an attempt for finding the international market.

(a)

(b)

Figure 44. (a) Packets of betel vine in market (b) Betel vine market busy for marketing.

2.11. Crop Rotation

Crop rotation is the practice of growing a series of dissimilar or different types of crops in the same area in sequential seasons primarily in order to balance the soil fertility and prevent the disease and pest. It provides various nutrients to the soil. It is one component of polyculture and can also improve soil structure by alternating deep-rooted and shallow-rooted plants.

In betel vine farming, crop rotation is also beneficial for preventing the disease and pest control as well as for high yield of leaf crop. Though the average life span of the betel vine is 10 to 12 years, the old or aged vines produced low quality leaf, showed low growth and are much susceptible by various diseases. Therefore, after betel vine cultivation, other types of crop can be grown in the same land for 1 or 2 years to enhance growth and protect the betel vine from disease infection. In some states of India, the farmers cultivate the various types of vegetables and cereals after 3 to 4 years betel vine cropping. For example, after about 2 years cultivation of Maghai betel vine cultivars, the farmers used the same land for vegetable cultivation in Bihar state of India. The vegetables like - cabbage, potato, onion, tomato, chili, brinjal, pumpkin, lady finger, etc. and cereals like - paddy, wheat, corn, various pulses, pea nut, cashew nut, ground nut, gram, etc. can be cultivated as intercrops in crop rotation practice. It is also found that the betel vine cultivation is continued for about 10 to 20 years without following the crop rotation process in some places of India and Bangladesh.

References

Chandra, G. and Sagar, R. L. (2004). Harvesting green gold: cultivation of betelvine in Sundarban. *Indian Farmers Digest* 37(3): 5-13.

Das, J. N., Das, S. C. and Acharya, A. (1989). Effect of nitrogen on yield and incidence of diseases in betelvine, Orissa. *Orissa J. Agric. Res.* 2: 177-179.

Mandal, B. K., Choudhuri, S. K., Sengupta, K. and Dasgupta, B. (1994). Response of betelvine varieties to organic manure and prilled urea. *Indian J. Agric. Sci.* 64: 297-301.

Rahman, O., Babu, S. H. C. and Rao, S. N. (1990). *J. Agric. Sci.* 100: 251-253.

Thomas, U. C., Chandini, S. and Thomas. A. (2013). Integrated Nutrient Management Studies in betelvine *(Piper betel L.). International Journal of Applied Research and Studies*, 2(7).

Thomas, U. C., Chandini, S., Anil Kumar, S., Thomas, A. and Satyanarayana, T. (2010). Evaluation of Different Nutrient Management Options for Leaf Yield, Quality, and Economics of Betel Vine Cultivation. *Better Crops – South Asia,* 16-17 (http://www.ipni.net/publication/bca.nsf/0/B553675FA4561866852579A4007A4B4B/$FILE/BCSA%202010%20pg%2016-17.pdf).

Chapter 9

PESTS AND DISEASE

1. INTRODUCTION

On account of immense medicinal, social, religious and export values, betel vine is recognized as an important crash crop intensively and extensively growing in large scale in different parts of the world. During cultivation, betel vine is very much affected by various pests and diseases those result in great loss to the farmers. Because, it occurs in a very virulent form and if not controlled, causes widespread damage and even total destruction of the entire betel vine farm.

Furthermore, the leaves are specifically affected by microbial pathogens onto the leaf surface by common agents - wind, soil and water. The most important diseases of betel vine plants are powdery mildew disease, leaf rot disease, foot rot disease and leaf spot disease. The leaves, which are subjected to these diseases, are unused resulting in tremendous economic loss to distributors, shop keepers and people involved in this trade (CSIR, 1969; 1984; Chaurasia, 2002). Some diseases occur in a very powerful form and cause unlimited damage and even entire demolition of farm (Vijayakumar and Arumugam, 2011).

Figure 45. Farmers taking care and precaution of betel vine farm.

It is apparent that betel vine yard is commonly threatened by various pest attack and disease infection (Sengupta et al., 2011; Jane et al., 2014). Therefore, farmers and practitioner associated with betel vine cultivation should regularly take care and remember that "Precaution is better that prevention and cure" following the ecofriendly farming practice for sustainable production (Figure 45).

2. Major Pest, Disease and Control

The major pests and diseases of betel vine and its control measures are described as below:

2.1. Pests

Although, no economically important pest problems are reported, but damages of the betel vine by sap sucking insect, and red mites are commonly found in betel cultivations. There are more than fifteen pests of insect attack and damage the betel vine yard. The symptoms of attack and control measures of some major insects are described as below:

Figure 46. Photograph of betel vine bug.

2.1.1. Betel Vine Bug (Dispunctus politus) (Figure 46)

Symptoms: The nymphs and adults damage the leaves by puncturing and sucking the juice causing the leaves to shrivel, fade and dry up.

Control: The pest can be controlled by spraying 0.05% Malathion.

2.1.2. Linear Scale Pest (Lepidosaphes cornutus)

Symptoms: It infects the leaves, petiole and main vines, which resulted in 30 to 35% yield loss. The scale insects are either light brown or dark brown in colour. Both the nymphs and adults suck the sap and the infested leaves loose colour, exhibit waxy, watery appearance, crinkle and ultimately dry up in most of the cases of severe. This kind of damage is generally found in old gardens.

Control:

1. Scale-free cuttings/setts should be selected for planting
2. Remove and destroy affected vines.
3. Spray either 5% neem (*Azadirachta indica*) seed kernel extract or 2% neem oil emulsion or chloropyriphos 20 EC at the rate of 2 ml/l or Spray NSKE 5% or Malathion 50 EC 1 ml/l of water and repeat after 21 days if necessary. When one or two scales are noticed on the basal portion of the stem/leaves, direct the spray solution to the basal portion of the vines.

Figure 47. Photograph of mealy bug.

2.1.3. Mealy bug (Ferrisia virgata) (Figure 47)

Symptoms: The leaves are covered with white masses, and the insects suck the sap from leaves and shoots.

Control: Mealy bugs can be controlled by spraying of Chlorpyriphos 20 EC at 2 ml/l or Dimethoate 30 EC 2 ml/l or Malathion 0.05%. Concentrate the spray towards the collar region.

2.1.4. Red Spider Mite (Tetranichus Sp.)

Symptoms: Found in the lower surface of leaves and suck the sap of the leaves (Figure 48). The leaves become pale yellow with numerous small spots which losses the quality of the leaves.

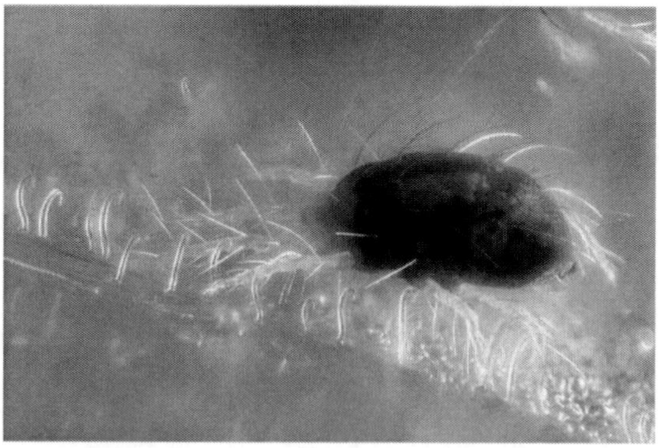

Figure 48. Photograph of red spider mite.

Control: Mites can be effectively controlled by spraying of 0.03% sulphex or wettable sulphur 50 WP @ 1 g/l or Dicofol 18.5 EC 0.5 ml/l.

2.1.5. Nematodes

Three types of nematodes viz. Root knot nematode (*Meloidegyne incongnita*), Reniform nematode (*Rotylenchus runiformis*) and Lance nematode (*Haplolamus indicus*) attack to the betel vine yard.

Symptom: The root-knot nematodes affected plants become stunted in appearance bearing yellowish coloured small-sized leaves and swollen aerial roots in the cluster. When they are uprooted, deformed roots with prominent galls and swelling are observed. They are also found frequently associated with different stages of rotting caused by bacterial and fungal infection. Eventually, the affected plants wilt and die. Reniform nematode frequently associated with betel vine roots and generally causes unthrifty growth of plants producing small feather like leaves. The Lance nematode produced stunted growth of plants with general yellowing of leaves.

Control: For effective control of nematodes, following practices can be applied:

1. Chopped shade dried *Calotropis* leaves and neem @ 2.5 t/ha can be applied to soil at the time of planting.
2. Oil cake of neem or mustard @ 500 Kg/ha to 2 t/ha.
3. Carbofuran 1.5 Kg per hectare in fresh planting in endemic areas two weeks prior to planting to manage nematodes in soil.
4. Soil application of *Bacillus subtilis* (BbV 57) or *Pseudomonas fluorescens* @ 10 g/vine for the control of root knot nematode and quickly wilt of betel vine.

2.1.6. Aphids (Aphis gossypii)

Symptom: Nymphs and adults suck the sap of leaves and tender shoots (Figure 49). Black shooty mould develops on the leaves during the end of January to March.

(a) (b)

Figure 49. Aphid (a) and damage of betel leaf caused by aphid (b)

Control: It can be effectively controlled by spraying 0.03% Metasystox or Chlorpyriphos at 2 ml/l on Agathi leaves. Clip off excess Agathi leaves.

2.1.7. White fly (Dialeurodes pallida) and Black Fly (Aleurocanthus rugosa)

Symptom: White and black flies suck the sap of leaves resulting in the development of discolouration of patches and yellowish marking appear on infected leaves considerably (Figure 50).

Figure 50. White and black flies.

Control: This can be effectively controlled by applying:

1. Confidor 200 SL or Neurocombin @ 1-2 ml/litre water (0.2%)
2. Leaf eating caterpillar can easily be controlled by spraying malathion at the rate of 0.5%.
3. Before planting setts, Heptachlor 37 kg/ha as a basal dose should be applied to the soil in order to take a precaution from damage.

2.2. Disease

Some fungal and bacterial pathogens are found to attack the betel vine which is responsible to cause severe damage and devastation of entire betel vine field. The pathogens, symptoms and control measures are described in details as below:

2.2.1. Bacterial Disease

Leaf Spot and Blight Spot

It is a bacterial disease caused by *Xanthomonas compestris var. Betlicola* showed a decrease in moisture, chlorophyll, protein, phenol and total carbohydrate contents as compared with healthy leaves (Wasnikar et al., 1993). This is the most serious disease affecting the betel vine yard. Disease becomes epidemic during rainy seasons. First characteristic symptom is moist oily patches on underneath of leaves. It gradually spreads and turns into brown or black in colour (Figure 51). Later, these appear on the upper surface also as dark round angular ones surrounded by yellowish zone. When the condition is serious, these patches can spread to the stem resulting of shedding leaves and nodes. Consequently, the plant will die but the disease can easily spread into surrounding vines.

Control: High precautions are most important and only way to reduce the incidence of this disease in the vineyard.

Figure 51. Bacterial leaf spot disease caused by *Xanthomonas compestris var. Betlicola* (https://colloque4.inra.fr/var/epidemiology_canopy_architecture/storage/fckeditor/file/6_Betelvine-DAS-2007.pdf).

- Two to three sprays of 0.25% Zineb or Zirum should be made after every six months as a protective measure.
- The disease intensity can also be reduced by regulating the shade in the vineyard.
- Drenching with wettable Ceresan 0.1% or Agallol 0.1%
- This disease can effectively be controlled during primary stage by spraying either of 1% Bordeoux Mixture and Streptomycin Sulphate 250 to 500 ppm or 0.25 to 0.3% Zineb and Streptomycin Sulphate 250 to 500 ppm at a monthly interval.
- Uproot the diseased plants and burn or bury them to protect other vines.
- Never use an infected cutting for planting in the new vineyard.
- Spray of plantomycin 500 ppm is also useful in controlling leaf spot.

2.2.2. Fungal Disease

Foot Rot or Leaf Rot

Fungal foot rot and leaf rot attacks on basal stem and leaves, respectively, are other major disease problems in betel cultivation. The disease is caused by soil borne pathogen, *Phytophthora parasitica var. piperina* Dast. (Figure 52a) and *Sclerotium rolfsii* (Figure 52b), occurs in severe form in rainy season. The first symptom of the foot rot is darkening of the basal stem (Figure 52c,d). Leaf rot is caused by *Phytophthora parasitica* (Figure 53). Soon the vine wilts, accompanying by yellowing of leaves and drooping of the vines, starting from the tip to downwards. The first symptom of leaf rot is the appearance of water soaked lesions (Figure 53). The infected spots enlarge rapidly to cover the blade, which starts rotting. Diseased leaves turn brown to dark brown and later dirty black.

Control:

i. At the time of planting, the cutting/setts should be dipped at least for 20 minutes in 1% bordeoux mixture or 0.1% ceresin or 30 minutes in Streptocycline 500 ppm.
ii. Bordeoux mixture (0.5%) or 0.2% Zineb (Dithane Z 78) should be spread on the ground near vines and also on diseased leaves at an interval of 15 days from last week of June to October.
iii. Once the plant is diseased, it should be removed and partially soil must be sterilized by drenching with chestnut compound or adding lime and mix thoroughly.

Integrated method for the management of *Phytophthora* wilt:

i. Select well matured (more than 1 year old) seed vines free from pest and diseases.
ii. Apply 150 kg N/ha/year through neem (*Azadirachta indica*) cake (75 kg N) and Urea (75 kg N) and 100 kg P_2O_5 through Super phosphate and 30 kg Muriate of potash in 3 split doses first at 15 days after lifting the vines and second and third dose at 40 - 45 days intervals. Apply on beds, shade dried neem leaf leaves at 2 t/ha and covers it with mud (2 t in 2 split doses).
iii. Drench Bordeaux mixture 0.25% in basins formed around the vine at monthly intervals, three times soil drenches and six times spray can be followed.

iv. During winter season, avoid frequent irrigation.
v. Remove the affected vines away from the garden and burn them.
vi. Application of *Trichoderma viride* at the rate of 5 g/vine.

Figure 52. Foot rot of betel vine caused by *Phytophthora parasiticaa* (a), *Sclerotium rolfsii* (b) (https://colloque4.inra.fr/var/epidemiology_canopy_ architecture/storage/ fckeditor/file/6_Betelvine-DAS-2007.pdf), and stem lesions (c and d) (Bhattacharya et al., 2012).

Figure 53. Leaf rot of the betel vine.

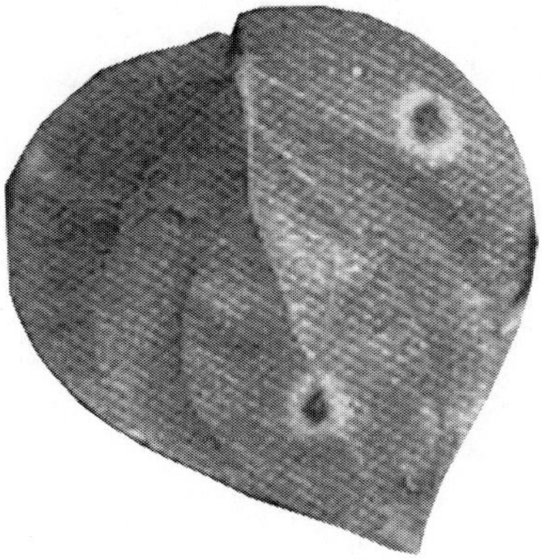

Figure 54. Betel leaf spot disease caused by fungi *Colletotrichum capsici.*

Anthracnose

The disease is caused by fungi *Colletotrichum capsici*, a set borne pathogen which is a very serious disease. Small black circular spot or lesions appear on leaves expanding rapidly in humid conditions. Leaf spots appear as a brownish black centre with yellowish hallow around and in severe cases the leaves drop owing to shrinkage of tissues (Figure 54).

Control:

1. Before planting the cutting/sets should be dipped in 0.1% Ceresan or other organo mercurial solutions.
2. Spraying once or twice 0.2% solution of Zineb (Dithane Z 78) or 0.1% Bavistine.

Powdery Mildew

It is a fungal disease caused by *Oidium piperina*. The disease affects the foliage, causing several damages. Small white patches appear on both the surfaces of young leaves. These patches enlarge running together to cover large portion of the blade, covering the leaf surface with characteristic whitish powdery coating (Figure 55). It is an air borne disease usually appears in the month of January and February in India. Warm humid weather and cold nights favour the spread of disease.

Figure 55. Front (a and b) and back view (c and d) of normal (a and c) and powdery mildew disease infected (b and d) betel vine leaves (Vijayakumar and Arumugam, 2011).

Control: Sulphur dusting, the most effective method of controlling powdery mildew disease, cannot be used because it renders the leaves unfit for chewing purposes. This disease can be controlled by spraying of Sulphex (0.25%) or Benlate (0.25%).

Betel Vine Wilt

In severe condition, plant started wilting and shown water stress condition caused by *Phytophthora capsice leomian.*

Control: It can effectively be controlled by applying neem leaf extract (2%) or neem cakes extract (0.5%).

2.2.3. Other Disease

Dew or Fog Injury

Betel vine leaves are also damaged due to dew or fog injury. In winter season, the leaves are injured due to drops of dew, which turns into white spots and blotch. This injury can reduce the market price of infected betel leaves.

Control:

1. The disease can effectively be controlled by spraying 0.3% Zineb (Dithane Z 78) or 0.25 to 0.3% Zirum spray 2-3 times at a fortnight interval.
2. The disease incidence can also be lowered through extra layering of upper surface of boroj by paddy straw.

References

Bhattacharya, R. Mondal, B., Ray, S. K. and Khatua, D. C. (2012). A study on bacterial disease of betelvine in West Bengal, India. *International Journal of Bio-resource and Stress Management,* 3(2): 211-216.

Chaurasia, J. P. (2002). Betelvine Cultivation and Management of Diseases. Scientific Publishers, India.

CSIR (Council of Scientific and Industrial Research, New Delhi) (1969). *The Wealth of India*, 8: 84-94.

CSIR (Council of Scientific and Industrial Research, New Delhi) (1984). *The Wealth of India,* 8: 87.

https://colloque4.inra.fr/var/epidemiology_canopy_architecture/storage/fckeditor/file/6_Betelvine-DAS-2007.pdf.

Jane, N. S., Deshmukh, A. P. and Joshi, M. S. (2014). Review of study of different diseases on betelvine Plant and control measure. *International Journal of Application or Innovation in Engineering & Management*, 3: 560-563.

Sengupta, D. K., Dasgupta, B., and Datta, P. (2011). Management of foot rot of betel vine (*Piper betle* L.) caused by *Phytophthora parasitica. Dastur Journal of Crop and Weed,* 7(2): 179-183.

Vijayakumar, J. and Arumugam, S. (2011). Foot rot disease identification for the betelvine plants using digital image processing. *Journal of Computing*, 3(2): 180-183.

Wasnikar, A. R., Khatik, S. K., Nayak, M. L., Vishwakarma, S. K. and Punekar, L. K. (1993). Effect of phosphorus on bacterial leaf spot disease incidence, and chemical composition and storage quality of *Piper betle* leaves. *Phytoparasitica*, 21: 75. doi:10.1007/BF02980874.

Chapter 10

MATERIALS AND TOOLS

1. INTRODUCTION

Although the betel vine is one of the most important economic crops in some countries due to its multipurpose applied significances, nonetheless, the cultivation practices have not been modernized so far. Therefore, it is still recognized as a troublesome and labourious farming, whereas plants need too intensive caring and nourishment. The highly profitable betel vine cultivation practice involves several steps, which may vary in different countries. However, the traditional cultivation practice uses various types of materials and tools in different steps of the farming, such as, construction of boroj, planting, lowering, soiling, watering, harvesting, etc. Recently, so many machines and tools have been adopted to facilitate the different steps of farming. For example, power tiller for soil tillage, shade net for shedding the betel garden (boroj), specific water pump with pipes and different accessories for watering the betel vine, etc.

2. Materials and Tools

The materials and tools are commonly used in different stages of cultivation practice are as follows:

2.1. Materials

There are several kinds of material generally used by betel vine farmers in the cultivation process depending on the specific area, availability, cost effective, durability of materials, etc. Farmers basically select the farming materials considering the above characteristics. Followings are the illustrations of some materials:

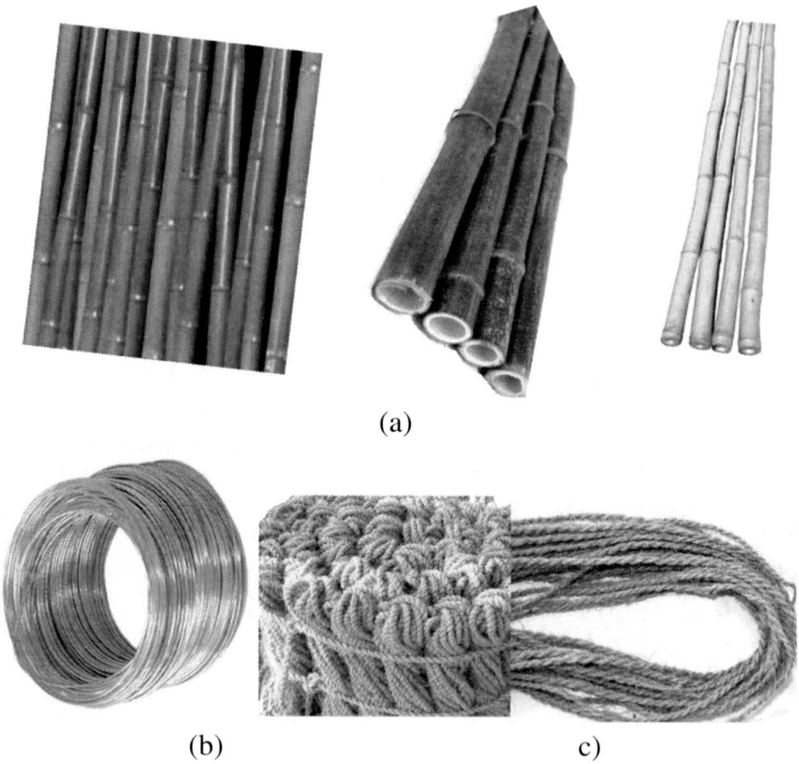

Figure 56. Bamboo pole (a), iron wire (b) and coconut thread (c) commonly used in betel vine garden (or Boroj) construction.

i. *Bamboo pole, iron wire, coconut thread, etc.* are the basic requirement for construction of betel vine garden or "boroj" (Figure 56a, b, c). These may vary in different places and in different countries. Some farmers use iron poles instead of bamboo.

ii. *Plant's leaves* such as coconut, date, palm, banana, etc. leaves are used as shade on the roof of the garden (boroj) (Figure 57 a-d). Now-a-days, shade nets produced from polyethylene have been developed and employed for this purpose (Figure 57e).

(a)

(b)

Figure 57. (Continued).

(c)

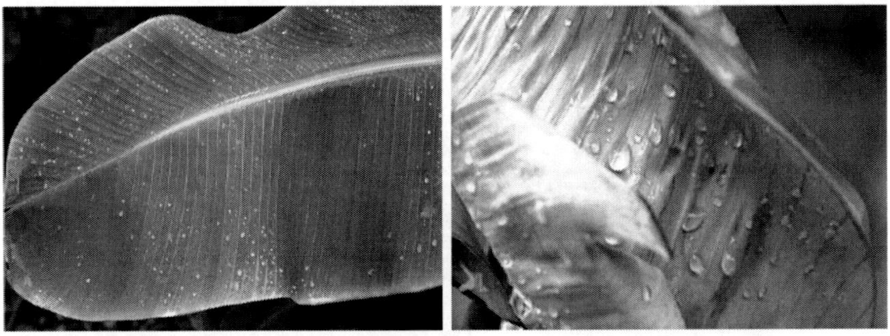

(d)

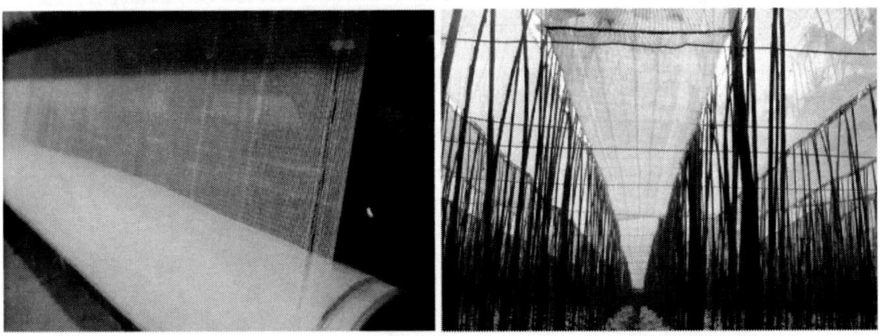

(e)

Figure 57. Coconut leaf (a), date leaf (b) palm leaf (c) banana leaf (d) [left panel and right panels resenting the green and dry respected leafs, respectively] and shade net (e) commonly used for making roof and boundary wall in the betel vine garden (Boroj).

iii. *Paddy straw or polythene sheet* is also sometimes employed for preparing the side walls of the boroj (Figure 58) in order to protect the betel vine from outside extreme environmental conditions, such as, heavy rain and wind, high temperature, heavy cold, etc.

(a) (b)

Figure 58. Paddy straw-left panel (a) and polythene sheet-right panel (b) used for building roof and boundary wall in the betel vine garden (boroj).

Figure 59. Thin threat like grass derived organic material used in binding the betel vine with supporting stick.

iv. *Thread like grass (organic binder) and stick or bamboo poles* are most important to bind the betel vine with climb stick in lowering process (Figure 59). Thin thread like grass is an organic binder which helps to bind the betel stem with the stick. The thin stick or bamboo pole is a supporter on which vine can creep or climb up.

2.2. Tools

Likewise materials, though tools used in this farming practice are varied place to place, some tools are very common applied in betel vine garden preparation and management as below:

i. *Power tiller and spade* are important tools for preparing soil in the newly constructed betel vine farm. It helps to tillage the soil for treatment. The spade is required for removing and tilling the soil by hand (Figure 60).

(a)

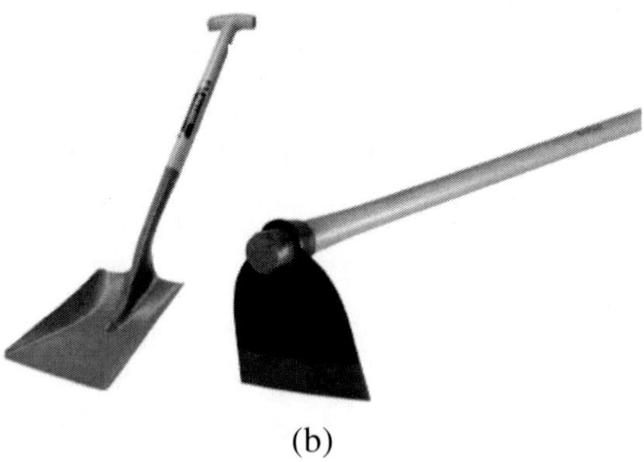

(b)

Figure 60. Power tiller - upper panel (a) and two kinds of a spade - lower panel (b).

Figure 61. Different kinds of ladder.

ii. *Ladder is* very much helpful in initial phase for constructing the frame of garden (boroj) and throughout the year for repairing the structure (Figure 61).
iii. *Knife* is required for cutting the large vine into small pieces favourably used for cutting preparation for plantation.
iv. *Soil carrier basket* is most important to carry out the betel vine earthing process after lowering. Soil carrier basket helps to carry the soil from outside to the inside of the boroj for covering the stem lowered (Figure 62).

(a) (b)

Figure 62. Soil carrier basket made by aluminium (a) or made by bamboo (b).

Figure 63. Water pump and pipe.

v. *Water pump machine and pipes* are required for irrigation and watering the farm of vine (Figure 63).
vi. *Spryer* is used for water spraying in newly planted cutting as well as for spraying the insecticides or pesticides or antimicrobial agents to treat diseases of the vine (Figure 64).
vii. *Artificial nail* is used for pick up the leaves from plants.

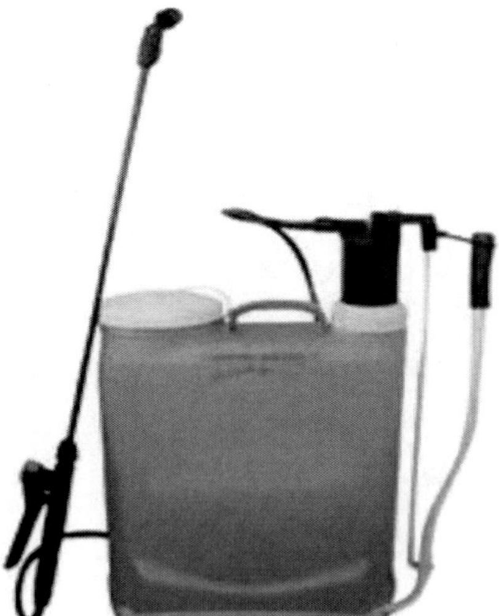

Figure 64. Common sprayer for spraying water, pesticides, fungicides, antipathogenics, etc. in betel vine farm.

Figure 65. Basket of bamboo is using in the betel vine collection in the garden.

viii. *Basket, commonly made by bamboo,* helps to harvest, arrange and carry the arranged leaves from one place to other places (Figure 65). Plastic basket or bucket is also used for the application of fertilizers.

Chapter 11

Production and Economy of Cultivation

1. Production

The betel vine is considered as one of the most profitable crops cultivated in many countries of the world. Though, the yield of leaves is varied by the influence of various factors, interestingly, the annual average yield of a healthy plant is about 60 to 70 leaves/plant/yr and 6 to10 million leaves/ha/yr are produced in a standard farm.

The leaves may also be retained on the vines for about six months without any visual signs and symptoms of deterioration if the vine farm is not infected by any disease. Long time retention of leaves in plants usually improves the quality of leaves as well as decreased the yield of leave. This provides an opportunity to escape the dull markets, but the total production may be affected.

The production of betel vine leaves is primarily depended on types of cultivar, soil qualities and various environmental conditions. The total production may also severely be affected by pests, bacterial and fungal diseases. Besides, the natural calamities such as, heavy rain, flood, strong wind, prolonged summer with high temperature, etc. sometimes may cause the immeasurable loss in yield of betel leaves. Additionally, spoilage of leaf after harvest may cause serious damage in benefit of farming.

Consequently, farmers suffer from a great economic loss. Recently, drying process has developed to protect such a leaf spoilage problem (Balasubramanian et al., 2011). However, qualitative and quantitative production of betel leaf is the important key for beneficial farming. A research work reported the yield of betel vine leaf in different places of Bangladesh from year 1995 to 2011 (Table 15) (Islam et al., 2015).

2. Economy

2.1. Production Cost and Income

The betel vine plays a significant role in the social, cultural and economic aspects not only in India, but also in several other countries of the world due to having great demand of leaves. It is commercially cultivated in countries such as India, Bangladesh, Pakistan, Malaysia, Indonesia, Sri Lamka, Thailand, Papua New Guinea, Madagascar, Bourbon and West Indies. India is the major producer; where it is cultivated on an area of about 55000 hector with an annual production worth about Rupees 9000 million (Table 16). In India, leaves worth about Rupees 30 to 40 million are exported to the countries like Bahrain, Canada, Great Britain, Hong Kong, Italy, Kuwait, Nepal, Pakistan, Saudi Arab and many other European countries (Jana, 1996; Singh et al., 1990). Bangladesh is the second largest producer. In India, the crop is extensively cultivated in the states of Andhra Pradesh, Bihar, Gujrat, Karnataka, Kerala, Maharashtra, Madhya Pradesh, Assam, Orissa, Rajasthan, Tamil Nadu, Uttar Pradesh and West Bengal. Guha (2006) reported that about 400000 to 500000 agricultural families are engaged in this farming practice in India.

As a consequence of high international demand, leaves worth of about Rupees 30 to 40 million are exported to the other countries such as Oman, Kuwait, Quatar, Saudi Arabia, UAE, UK, USA and Nepal are the main importers. Thus, in India, it is the most promising commercial leafy crop capable of attracting substantial amount of foreign exchange to the country. Revenue generated would easily exceed if agronomic practices are scientifically explored and improved.

Table 15. Area and production of betel leaf in Bangladesh (Islam et al., 2015)

Year	Area (ha)	Production (tons)	Yield (t/ha)
1995-96	13943	71910	5.16
1996-97	14595	77035	5.28
1997-98	14832	79080	5.33
1998-99	13820	73525	5.32
1999-00	15063	78780	5.23
2000-01	15346	82260	5.36
2001-02	14696	80540	5.48
2002-03	15472	83830	5.42
2003-04	16480	93425	5.67
2004-05	16771	93820	5.59
2005-06	16275	97415	5.99
2006-07	16536	101240	6.12
2007-08	17346	97947	5.65
2008-09	17643	105448	5.98
2009-10	17871	91681	5.13
2010-11	18247	105953	5.81
Mean	15934	88368	5.53
CV (%)	8.84	12.81	5.63
Growth rate (%)	1.80	2.50	0.70

The economics of cultivation were evaluated both in terms of benefit-to-cost ratio and net income. The expenditure for initial crop establishment was higher in the first year than in the second; as was the cost of cultivation and these factors were reflected in generally lower (Thomas et al., 2013). The initial cost of cultivation of betel vine including construction of a boroj may be about Rupees 100000 to 200000/ha at the minimum during the first year that may come down to about Rs 50000 to 60000/ha in the subsequent years and a minimum net profit of Rs. 50000 to 100000/ha/year or more up to Rupees 500000/ha/year is not an unexpected value from a well-established farm though the figures may vary due to several factors like agro-climate, location of the farm, variety, demand and supply, season, variation in price of the leaves (which sometimes may vary significantly in the evening compared to morning), inflation, etc. Such, wild and unwarranted fluctuation in price of betel leaves may discourage cultivation of the crop and endanger the economic stability of the farmers (ICAR, 2000; Guha, 2006).

However, a picture of average cost for establishment and maintenance of betel vine garden and average annual income has been drawn in the following tables (Table 17a, b) by researchers on the basis of information directly gained from expert farmers who are regularly engaged in betel vine farming in the practical field. It is also clear from the tables that the benefit of betel vine cultivation largely depends on various factors; especially labour along with some other ingredients whose values are variables at different times. The most important factor for annual income is market demand. The farmers get good or appreciable return when market demand of the betel vine is high.

Table 16. Estimated area under betel vine cultivation in major states of India (Balasubrahamanyam, 1994)

SN	Name of the States	Area (ha)	Area (%)
1	Andhra Pradesh	2900	5.80
2	Tamil Nadu	5500	11.00
3	Karnataka	8700	17.40
4	Kerala	3300	6.60
5	Orissa	5000	10.84
6	Gujarat	200	0.40
7	Maharashtra	2700	5.40
8	Madhya Pradesh	1250	2.50
9	Rajasthan	50	0.10
10	Uttar Pradesh	2000	4.00
11	Bihar	3200	6.42
12	West Bengal	3000	6.00
13	Assam	3000	6.00
14	Others	9200	17.56
	Total	50000	100.00

Table 17. An analysis of average expenditure and annual income of betel vine cultivation for the year 2000 and 2010 (http://www.downtoearth.org.in/coverage/too-costly-to-grow-33173) (a) and year 2016 (b) (Bhakta et al., 2016)

Table 17. (a)

Items (10 decimal or 0.04 ha)	**Year 2000 (Cost in Rupees, Rs.)**	**Year 2010 (Cost in Rupees, Rs.)**
Cost of setting up a betel garden		
Soil for elevation (2 ft high from ground to prevent submergence)	500	2000
Bamboo and other grass/straw for the outside structure	5500	17500
Organic fertilizers and pesticides	500	1000
Chemical fertilizers and Pesticides	2500	5000
Stem cuttings of betel vine for plant propagation	4500	15000
Stick on which the plant climbs	2000	4500
Pump set for water supply from nearby sources	7000	15000
Labour costs	4000	15000
Total initial investment	26500	75000
Running and maintenance cost		
Climbing sticks replaced every 3 months	1500	4000
Fertilizer, pesticides, fungicides and hormones, etc.	500	1000
Repairing outside structure	300	1000
Labour costs every 3 months (2 labours twice a week @ Rs. 150 per day)	2000	7500
Diesel cost for three months	400	1000
Total cost of maintenance	4700	14500
Average annual income		
First harvest in 3 months	1500 – 3000	2000 – 4000
Subsequently, every 2 weeks	1500 – 3000	2000 – 4000
Average annual income	36000 – 72000	48000 – 96000

Note: All calculations done according to farmers in Jaleswar, Balasore districts, Orissa, India

Table 17. (b)

Items (10 decimal or 0.04 ha)	**Year 2016**	
	Cost in Rupees (Rs.)	**Cost in US $ (@ 1 US $= Rs. 66)**
A. Cost of setting up a betel garden		
Soil for elevation (2 ft high from ground to prevent submergence)	5000	75.75
Bamboo and other grass/straw for the outside structure	30000	454.54
Organic fertilizers	3000	45.45
Chemical fertilizers and Pesticides	3000	45.45
Stem cuttings of betel vine for plant propagation	10000	151. 51
Stick on which the plant climbs	7000	106.06
Pump set for water supply from nearby sources	20000	303.03
Labour costs	30000	454.54
Total cost	108000	1636.36
B. Running and maintenance cost		
Climbing sticks replacement	8000	121.21
Fertilizer, pesticides, fungicides and hormones, etc.	7000	106.06
Repairing outside structure	8000	121.21
Labour costs every year	30000	454.54
Diesel cost for pump	3000	45.45
Total cost	56000	848.48
C. Average annual income		
1st year First harvest in 3 months Subsequently every 2 weeks (9 months)	 15000 – 20000 3500 (63000) – 5500 (99000)	 227.27 – 303.03 53.03 (954.54) – 83.33 (1500)
Total income	78000 – 119000	1181.81 – 1803.03
Subsequent years every 2 weeks	3500 – 5500	53.03 – 83.33
Total income	84000 – 132000	1272.72 – 2000

Note: All calculations done according to farmers in Medinipur district, West Bengal, India

From the above analysis of expenditure and income, it is apparent that betel vine is a suitable cash crop for farmers, despite high input requirement. Farmer can set up a betel vine garden only on a small about three decimal land area. From practical experience and also from investigation of some scientist, it has been established that a farm of 10 to15 decimals can provide considerably good net profit for a family of five persons during the period from 10 to 30 years. Therefore, the betel vine farm is recognized as "household bank". Since, it can provide cash to a farmer at every 15 days interval.

2.2. Impact of Betel Leaves on Economy

The betel vine is grown as an important cash crop, because, its leaf has promising and multipurpose beneficial properties capable of attracting substantial amount of foreign exchange to the country. It has estimated that about Rupees 10000 millions are the annual turnover for betel vine with substantial foreign exchange through export (Kumar, 1999). Presently, the betel vine cultivation recognized as a small cottage industry due to having vast economic potentiality of the crop. It can be adequately established by the fact that about 15 to 20 million people consume betel leaves in India regularly (Jana, 1996) besides those in other countries of the world which may include over 2 billion consumers (Jeng et al., 2002). That apart a small Boroj of about three decimal area can generate employment opportuniies for an agricultural worker throughout the year (Bhowmick, 1997) helping him to maintain his family. Further, as far as the national employment generation is concerned, it is estimated that about 20 million people derive their livelihood directly or indirectly, partly or fully from production, processing, handling, transportation and marketing of betel leaves in India, which includes about 5 million workers from West Bengal (Jana, 1995; Jana, 1996). In this way, the crop provides a national income to the tune of Rupees 6000 to 7000 million every year, and at the same time it also provides an income of Rupees 800 to 1000 million to the state of West Bengal. In addition to this, the railways earn about Rupees 100 million every year from transporting betel leaves from West Bengal to different parts of India like Bilaspur, Cochin, Gondia, Gandhinagar, Hyderabad, Jabalpur, Jalgaon, Katni, Nagpur, New Delhi, Raipur, Srinagar, Trivendram, etc. (Jana, 1995; Samanta, 1994). Presently, betel vine is extensively cultivated in India in almost all the states including Punjab, Haryana, Himachal Pradesh and Jammu and Kashmir.

The leaves are also in great demand in several other countries of the world where it is either not grown at all or the demand exceeds the local supply. Therefore, it has good export potential, and India exports huge betel leaves to the other countries. Consequently, leaves worth about Rupees 30 to 40 million are exported to the countries like Bahrain, Canada, Great Britain, Hong Kong, Italy, Kuwait, Nepal, Pakistan, Saudi Arab, many other European countries and USA (Jana, 1996; Singh et al., 1990). This clearly indicates the foreign exchange earning potentiality of the crop, which is required to be strengthened in the interest of the nation. If merely transportation and marketing facilities, including the export channels were developed adequately, then the revenue generated by this leafy crop would easily exceed that generated by any major crop of the country even with the present level of traditional agronomical practices. In fact, the revenue generated by the crop may be further magnified by many folds if the agronomic practices are scientifically explored.

However, it is the most important economic crop which uplifts the socio-economic status of certain percentage people. It has the immense potential to generate huge employment, especially in remote rural sector. It altogether significantly justifies its nomenclature as the "green gold of nature". A well-coordinated effort by the farmers, traders, scientists, administrators and policy makers is required to be initiated to boost up the national economy through proper exploitation of this green gold in multidimensional perspectives.

REFERENCES

Balasubrahamanyam, V. R. (Ed.). (1994). Betel vine. National Botanical Research Institute, Lucknow, pp. 6-7.

Balasubramanian, S., Sharma, R., Gupta, R. K. and Patil, R. T. (2011). Validation of drying models and rehydration characteristics of betel (*Piper betel* L.) leaves. *Journal of Food Science and Technology,* 48(6): 685-691.

Bhakta, B., Bhakta, U. Ghanti, S., Jana, B. B., Ghanti, P., Yamane, S., Bhakta, J. N. (2016) Betelvine: A green gold for sustainable economic upliftment in rural India. *Int. J. Env. & Tech. Sc.* 3:76-82.

Bhowmick, S. (1997). Paan: Anadrita Laxmi (In Bengali). [Betel leaf: The Neglected Goddess of Wealth]. Moyna Prakashani, Calcutta, West Bengal, India.

Guha, P. (2006). Betel leaf: the neglected green gold of India. *Journal of Human Ecology,* 19(2): 87-93.

http://www.downtoearth.org.in/coverage/too-costly-to-grow-33173.

ICAR (Indian Council of Agricultural Research, New Delhi) (2000). Annual Report of All India Coordinated Research Project on Betelvine (ICAR). National Research Centre For Medicinal And Aromatic Plants, Boriavi, Gujarat, India.

Islam, Q. M. S., Matin, M. A., Rashid, M. A., Hoq, M. S. and Moniruzzaman (2015). Profitibility of betel vine leaf (*Piper betle* L.) cultivation in some selected sites of Bangladesh. *Bangladesh J. Agril. Res.* 40(3): 409-420.

Jana, B. L. (1995). Gram Banglar Arthakari Phasal-Paan (In Bengali). [Betel Leaf: A Cash Crop of Villages of Bengal]. Calcutta, India.

Jana, B. L. (1996). Improved technology for betel leaf cultivation. A paper presented in the *"Seminar-cum-Workshop on Betel leaf Marketing",* held at State cashew nut farm, Directorate of Agricultural Marketing, Digha, Midnapur, West Bengal, India.

Jeng, J. H., Chen, S. Y., Liar, C. H., Tung, Y. Y., Lin, B. R., Hahn, L. J. and Chang, M.C. (2002). Modulation of platelet aggregation by areca nut and betel leaf ingredients: Roles of relative oxygen species and cyclooxygenase. *Free Radical Biologyand Medicine,* 32(9): 860-871.

Kumar, N. (1999). Betelvine (*Piper betle* L.) cultivation: a unique case of plant establishment under anthropogenically regulated microclimatic conditions. *Indian journal of History,* 34(1): 231-239.

Samanta, C. (1994). Paan chaser samasyabali-o-samadhan: Ekti samikkha (In Bengali) [A Report on the Problems and Solutions of Betel Vine Cultivation]. In: A booklet Adhikari, H. R., Eds. Calcutta, West Bengal, India.

Singh, K. K, Balasubrahamanyam, V. R. and Kochhar, V. K. (1990). Effect of different packing methods, temperature conditions, treatment with chemicals on the senescence and storage behaviour of betel (*Piper betle* L.) leaves. *Journal of Plantation Crops,* 18(1): 23-28.

Thomas, U. C., Chandini, S. and Thomas. A. (2013). Integrated Nutrient Management Studies in betelvine *(Piper betel L.). International Journal of Applied Research and Studies*, 2(7).

Chapter 12

INTEGRATED BETEL VINE FARMING AND ECOLOGICAL ENGINEERING

1. INTRODUCTION

Traditional farming approach is almost mono farming based, i.e., farming of one kind of crop or animal in one field, specifically in rural farming area except very few uncountable commercial organizations those farming a number of crops or animals in integrated approach. This mono farming practice is not only low production rate and cost-benefit because of high production cost caused by various environmental constrains, but it is also unsustainable and poses hazardous environmental problems. Therefore, mono farming can be recognized as unsustainable farming approach, which is incapable to uplift the rural socio-economic status of farmers. Consequently, mono crop farming is a high degree uncertainty farming in respect to generate high benefit and employment, i.e., unsustainable. As a result, socio-economic status of the farmers is remained unthinkably very poor and majority of farmers maintaining their status of below the poverty level even in 21st century. In this contest, it is imperative to apply the suitable *integrated farming strategy* instead of still widely followed low benefit mono farming practice for enhancing the production, augmenting the income and uplifting the socio-economic status of a farmer.

2. INTEGRATED FARMING

Integrated farming system (IFS) is a commonly and broadly used word to explain a more integrated approach to farming as compared to existing monoculture approaches. It generally refers to agricultural systems that integrate livestock and crop production and may sometimes be known as integrated biosystems.

In recent years, it has been further supported by the concept of an all-round development of agriculture, animal husbandry, fisheries and other sideline occupations. Although a integrated farming is economically and environmentally sound, the motivation for integration should be the national policy of diversification of production.

However, it is economically and environmentally sound farming approach that integrates the various farming enterprises *viz.,* cropping, animal husbandry, fishery, forestry, etc. which has great potentialities in the agricultural economy. These enterprises not only supplement the income of the farmers but also help in increasing the family labour employment. The advantages of IFS can be characterized by the following potential features:

- The integrated farming system approach introduces a change in the farming techniques for maximum production in the cropping pattern and takes care of optimal utilization of environmental resources.
- The farm wastes are better recycled for productive purposes in the integrated system.
- A judicious mix of agricultural enterprises like agriculture, horticulture, dairy, poultry, piggery, fishery, sericulture, apiculture, etc. suited to the given agro-climatic conditions and socio-

economic status of the farmers would bring livelihood and prosperity in the farming.

3. Ecological Engineering

In the early 1960s, ecological engineering emerged as a new concept in the environmental science. The term "ecological engineering" was introduced by Howard Odum and others in 1963, as utilizing natural energy sources as the predominant input to manipulate and control environmental systems. It takes several decades to develop and refine the exact definition of ecological engineering.

However, it is an emerging study of integrating ecology and engineering, concerned with the design, monitoring, construction of ecosystems, and human society, having a specific beneficial goal for both environment and human society. According to Mitsch (1996) and Odum (1989), ecological engineering is "the design of sustainable ecosystems intends to integrate human society with its natural environment for the benefit of both". According to Mitsch and Jørgensen (1989), the goals of ecological engineering are:

1. the restoration of ecosystems that have been substantially disturbed by human activities such as environmental pollution or land disturbance, and
2. the development of new sustainable ecosystems that have both human and ecological values.

Further, Mitsch and Jørgensen (1989) also summarized the following five key concepts of ecological engineering:

1. it is based on the self-designing capacity of ecosystems,
2. it can be a field test of ecological theory,
3. it relies on integrated system approaches,
4. it conserves non-renewable energy, and

5. it supports biological conservation.

Its implementation is still undergoing adjustment, and its broader recognition as a new paradigm is relatively recent. Constructed (artificial) island can be represented as an example of ecological engineering approach. It integrates both environment and human society for the conservation of environment and for the benefit of human society. On the basis of above key concepts, the betel vine cultivation system can be designed integrating other related ecosystems applying the principles of ecological engineering as "Integrated betel vine farm" for the benefit of both environment and human society.

4. Integrated Betel Vine Farming

In betel vine cultivation, the ecosystem can be designed by applying the integrated approach for the sustainable development of human and surrounding environment. In addition to betel vine, vegetable cultivation, horticulture and floriculture can be incorporated in unutilized inside or surrounding areas of the farm yard. Besides, water reservoir or pond can be used for fishery and bank of the water reservoir or pond is to be a convenient place for the purposes of poultry or piggery. These all integrated approaches are collectively formed a constructed ecosystem which functions in sustainable approach. Such as, water reservoir or pond receives excess nutrients from the betel yard (Boroj) through the runoff water, which helps to enhance the productivity of ponds responsible for fish growth. The pig or poultry droppings can be used as manures for betel cultivation as well as fishery. The pond's bank may also be used for vegetable cultivation and pond water for aquaculture (Chandra and Sagar, 2004). Such, nutrients or energy can be recycled in this constructed ecosystem (Figure 66).

The integrated betel vine farming can be characterized more specifically as follows: the betel vine is considered as main agricultural crop along with other supporting agricultural crops (such as, vegetables, areca nut, coconut, fruits, etc.) and dairy, piggery, poultry, fishery, etc. are the appropriate livestock subsystems in this regard. (Figure 66) The waste products (such as, leaf) of betel vine can be used for animal food and farm manures of the livestocks (dairy, poultry, fishery, etc.) are the primary source of organic fertilizer for the cultivation of the betel vine and associated supporting crops used as subsystems in the integrated betel vine farming system. Thus, waste products of one system are effectively used as nutrient resource of other systems through recycling process in the integrated betel vine farming.

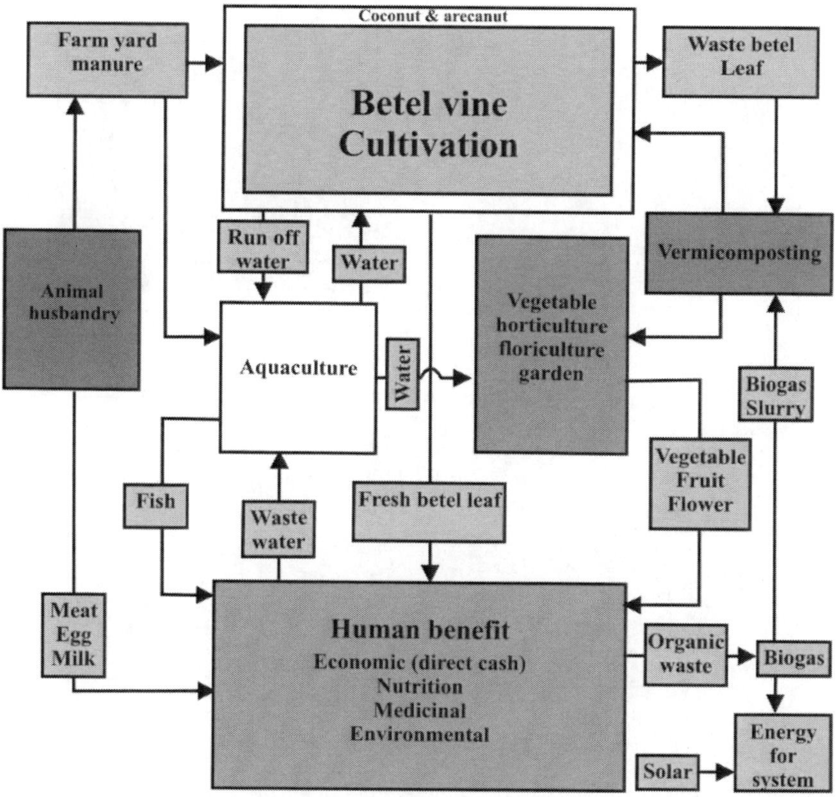

Figure 66. A model approach of integrated betel vine farming system exhibiting the integration of different components in system for recycling and reusing the products of one subsystem in other relevant systems. Arrow indicating the flow of product (i.e., nutrient or energy).

Figure 67. Original photos snapped from practical fields depicting different crops integrated in integrated betel vine farming (a) Betel vine cum fishery cum banana plant farming, (b) Betel vine cum fishery cum coconut tree farming (*Cocos nucifera*), (c) Betel vine cum fishery cum areca nut (*Areca catechu*) farming, (d) Betel vine cum elephant yam (*Amorphophallus paeoniifolius*) farming, (e) Betel vine cum turmeric plant (*Curcuma longa*) farming.

In betel vine cultivation, many integrated farming processes are practiced in different parts of India, Bangladesh and Nepal. Fishery, poultry, vegetable cultivation, horticulture and floriculture, etc. are commonly integrated as supporting crops with the major crop betel vine (Figure 67). This integrated cultivation of the betel vine is designed by following the sustainable approach for the benefit of the constructed artificial ecosystem and human society. It also provides extra benefits to the farmers along with the major cash crop of betel vine cultivation. The betel vine can be cultivated using the organic wastes scientifically, and thus it takes part in removing substantial amounts of greenhouse-gas carbon dioxide from the environment. In this view point, it plays significant roles in recycling the waste materials of the environment for the production of an economically valuable plant biomass. For this why, the integrated betel vine farming can be recognized as an ecological engineering approach for sustainable environment.

REFERENCES

Chandra, G. and Sagar, R. L. (2004). Harvesting green gold: cultivation of betelvine in Sundarban. *Indian Farmers Digest,* 37(3): 5-13.

Mitsch, W. J. (1996). Ecological Engineering: a new paradigm for engineers and ecologists. In: Schulze, P. C. (Ed.), *Engineering Within Ecological Constraints*. National Academy Press, Washington, D.C., pp. 114-132.

Odum, H. T. (1989). Ecological Engineering and Self-Organization. In: Mitsch, W. J. and Jørgensen, S. E. (Eds.), *Ecological Engineering: An Introduction to Ecotechnology*. John Wiley & Sons, New York, pp. 79-101.

Mitsch, W. J., and Jorgensen, S. E. (1989). "Introduction to Ecological Engineering." In: W.J. Mitsch and S.E. Jorgensen (Editors), *Ecological Engineering: An Introduction to Ecotechnology*. John Wiley & Sons, New York, pp. 3-12.

Chapter 13

ADVANCEMENT IN CULTIVATION: PROBLEMS AND SCOPE OF RESEARCH

1. INTRODUCTION

Betel vine is profitable crop cultivated in many Asian countries. The cultivation practices followed are very conventional, and any scientific refinement over the existing cultivation methods could not be advocated so far, due to lack of enough research findings under the agro-climatic situations in the leading tracts (Thomas et al. 2013). Although, betel vine has potential medicinal, herbal, ayurvedic values as well as significant direct economic importance, nonetheless, it is still neglected by researchers. It is also important to mention herein that although cultivation of the betel vine began long years ago (about 1,500 years ago in India), no such improvement has been found so far neither in the cultivation technique, nor in its industrial uses. However, a substantial research is needed for systematic genetic characterization of the cultivars or landraces in order to their conservation, variety-wise biochemical characterization and improvement of qualitative and quantitative properties of betel for the high benefit. The farming is facing lots of problems from production to marketing. Recently, the lack of advanced cultivation technology, leaf spoilage and storage, disease, high labour cost and crisis and marketing are recognized as most important problems in the betel vine farming. Therefore, there are lots of scopes and opportunities to carryout the research work in different steps of cultivation in order to develop novel methods, instruments and finding appropriate materials required for the betel vine farming. To develop the simple and novel cultivation technology, the specific research centre for the betel vine cultivation is essential.

2. Characterization of Btel Vine

A systematic genetic characterization of all cultivars or varieties of the betel vine is essentially important for their conservation, propagation and cultivation. Modern DNA fingerprinting technologies, such as, polymerase chain reaction (PCR), denaturing gradient gel electrophoresis (DGGE) and DNA sequencing including second generation pyrosequencing could be used to identify the different varieties for their proper conservation as well as to overcome the synonym problem and their proper authentication. Similarly, an effort should be paid in biochemical characterization of betel vine to explore new useful active compounds taking the advantage of modern technologies and tools - high performance liquid chromatography-mass spectrometry (HPLC-MS), Gas Chromatography- mass spectrometry (GC-MS), nuclear magnetic resonance (NMR) spectroscopy, etc. Using these techniques, characterization of essential oils of betel vine can be made to explore its advance biomedical applications in industrial and commercial fields.

3. Cultivation Technology

Qualitative and quantitative improvements of betel vine leaf are the important aspects, which directly and indirectly regulate the biochemical quality of leaf, its market demand and farmer's benefit. Current researches has paid much attention in the complex interaction between the plants, soils, microbes and associated secondary marker metabolites and other factors responsible for plant health and growth. Therefore, researches should focus light in this regard to improve the quality and quantity in betel vine production.

Because of using the traditional technology in different steps from cultivation, marketing and preservation to improve quality, the betel vine farming is very troublesome and labourious so far. There is always a demand to develop the simple and advanced cultivation technology in respect to novel methods, instruments and finding required materials. For example, power tiller for soil tillage, shade nets for shedding the boroj, specific water pump for watering the betel vine, etc. are used as some advanced betel vine cultivation facilities. The new techniques, materials and tools can be developed in construction of boroj, planting, lowering, soiling, watering, harvesting, etc. to make easy for the cultivation process and low-cost. Huge natural resources are available in surrounding environment those can be properly utilized in different steps of the betel cultivation which helps to develop a low-cost and sustainable organic farming.

The various tools and materials are commonly used in different stages of cultivation practice are already discussed in the earlier chapter. The followings are some materials and tools can be improved to advance and ease or simplify the different cultivation steps by substantial research afford:

1. Construction of Boroj: Permanent brick or iron structure and long lasting shade of roof and side wall.
2. Planting of cutting: Specific cutter such for cutting the vine.
3. Lowering of the vine: Binder for binding vine with support stick, support stick, soil carrier between the rows, hand plough.
4. Earthing up: Manufacturing of the specific device to reduce the hard labour required for earthing the vine.

5. Watering of vines: Pump aided with the simple automated device with pipes, water spryer, etc. which can work by sensing the humidity of farming system.
6. Harvesting basket: Basket with a wheel that can easily move through the space between two rows.

The concerned scientist of agricultural engineering has vast scope to carryout research work for developing various tools and materials to simplify this hard laborious betel vine cultivation process.

4. Leaves Spoilage and Storage

The Betel leaf is a very perishable commodity and therefore, always subject to wastage by quick spoilage due to dehydration (Figure 68), fungal infection, dechlorophyllation etc. Due to lack of proper preservation technology, the leaves are quickly spoiled, which causes huge damage after harvest that is known as post harvest loss (Madan et al., 2014). Therefore, the betel leaf should be conserved properly to get rid of spoilage after harvest. This post harvest loss may ranges from 35% to 70% during transport and storage. Even if the most conservative estimates of 10% loss were considered that too would reveal a menacing loss of betel leaves worth about Rupees 900 million every year to the country. Not only this, particularly in the rainy season a large portion of the leaves remains unsold or sold at a throw away price. In view of the alarming losses, the different techniques such as, drying the leaves, controlling senescence by chemical treatments, manipulation of storage temperature, adopting better packaging materials and methods besides curing and bleaching of the leaves are attempted to minimize the wastage. Such wastage may also be minimized by extracting essential oil from the stranded or unsold leaves are it fresh or stale or dechlorophylled or even partially decayed.

Figure 68. Spoilage leaf of the betel vine.

In this respect, a study is needed in chlorophyllase activity for long-term storage and improvement of export potential of the bete lvine leaf (Das et al., 2016). This study can demonstrate what genetic and biochemical mechanisms responsible for transforming the green colour of leaf to yellow colour. Unfortunately, there are very few reports available on the preservation of betel leaves and hence more of such research inputs are required in this area (Madan et al., 2014). Balasubramanian et al. (2011) developed drying methods in order to prevent the betel leaf spoilage maintaining the quality. However, it is obvious that there is vast scope of research in above mentioned different techniques to control the leaves spoilage and storage. This concerning research will help the farmers to obtain much benefit.

However, in order to reduce the post harvest losses of betel leaves, coupling of ancient techniques of solar drying and depetiolation to the modern methods of preservation, including modern drying technologies, modifying surrounding atmosphere, advanced packaging technologies, etc., can be beneficial. In this regard, Madan et al. (2014) focused on some important areas to overcome concerned problems: (1) more of the scientific research and technology inputs are required in post-harvest losses, (2) a well coordinated effort by the farmers, traders, scientists, technologists, administrators and policy makers will boost the national economy and generate huge employment opportunities for the people, (3) waste and by-product utilization in the industries should be focused, and (4) government should take initiative in this area for funding various projects under competent scientists to explore appropriate technology.

5. Disease Intervention

Betel vine crops are more prone to disease, which is one of the vital problems of betel vine cultivation. Though lots of studies are carried out to control various diseases, but most of them are basically involved in the application of hazardous chemicals as disease treatment methods. Therefore, an eco-friendly, sustainable and biological disease controlling measure still has demand in this respect. The vigorous research concerning the organic based disease management is also an important aspect to protect the environmental pollution caused by applying the chemical based disease controlling agent.

Following recommendation and suggestion so far has been given to farmers.

1. Soil solarization of the bete lvine garden in the month of April-May.
2. The root exudate of marigold (*Tagetes* sp.) reduces the population of nematode, therefore, marigold should be planted before cultivation of the bete lvine.
3. Soil application of neem cake 20 q/ha in two spilt doses in the month of July and December.
4. Application of 1% Bordeaux mixture to be spread on soil if not solarized.
5. Dipping of cutting in 0.5% copper oxychloride or 1% Bordeaux mixture for 30 minutes prior to planting.
6. Application of Trichoderma as set treatment for 30 minutes before planting. Soil application of Trichoderma with FYM 10 Kg/ha in June and December in affected area.
7. Setts treatment of IBA or Ceredex (100 ppm) for rooting of cutting.
8. Inter cropping of Pan pimpri (*Piper longum*) in the betel vine should be avoided.

9. In betel vine, orchards for shade and support planting of Pangra (*Erythrina indica*) is suggested instead of Shevri (*Serdania aeqyptica*).
10. Drenching of 1% Bordeaux mixture 10 litre per 5 sq m in the month from July to December once in a month or after two months near root zone of the betel vine. One Kg Trichoderma and 9 Kg FYM kept moist for 8 days and applied through so il @100 g/sq m for control the wilt of the betel vine.
11. For the control of leaf spot of betel vine, spraying 25 g copper oxychloride or 20 g Dithane M-45 in 10 litre of water, two sprays at an ineterval of 15 days should be given.
12. To minimize the betel vine wilt, implement the drip irrigation system for water management.
13. Application of phorate 10 G @ 4 Kg/acre through soil for the management of nematode infestation.

6. High Labour Cost and Crisis

The betel vine farming is facing serious problems of labour crisis and high labour costs. Majority of rural people are migrating to towns and cities to earn much money compared to rural areas. Besides, the farming practices are hard and labourious in nature due to unavailability of convenient farming instruments and equipments, which ultimately resulting in serious crisis of appropriate labour in the rural betel farming. To compensate and overcome the problem of labour crisis, it has been found that female members of farmers are frequently and actively participated in different stages of farming practice (Figure 69). All these factors are significantly responsible for enhancing the labour costs which lead to reduce the farm benefit (Bhakta et al., 2016). In spite of that, the price of the betel vine should be high enough to employ the labour of high rate and get the high benefit. The betel vine farming can be benefited if above all factors are considered favourably.

(a) (b)

Figure 69. Female members of farmers are engaging in management of farm (a) and arrangement of leaves for marketing (b).

The research for developing the simple instrument for carrying out the hard and labourious betel vine farming in the simple process without hard and huge labour is inevitable in this context.

7. Marketing

The organized marketing systems of a crop may help to obtain the actual and high benefit (Bhakta et al. 2016). The marketing of betel leaves is one of the most important problems of farmers. The lack of organized market and proper transport systems in the production areas has a significant impact in marketing. The poor marketing and transport systems reduce the market price of betel leaves and benefit because of not marketing in proper time. In spite of that, the betel leaves are rotten during the period of transportation due to its perishable nature. These organized market and transportation facilities should be considered to avoid the loss of farming. The middle-men are one of the major problems in the marketing system. These middle-men interrupt the marketing process and enjoyed a lion part of farming benefit. Consequently, the farmers are suffered a lot to obtain the actual price or benefit from betel farming. Finding of new and demanding international market is also an important aspect to export huge amount of leaves, which may enhance the demand as well as price of betel leaves.

This may be safeguarded by evolving a well regulated organized marketing system for which research work, particularly on the marketing systems and intelligence and allied aspects relevant to the crop are required to be initiated.

Allthough several researches have been performed and developed useful products and solved many problems by a number of scientists, applied potential research and technological inputs are still required in this regard for advance and sustainable cultivation of the betel vine.

REFERENCES

Balasubramanian, S., Sharma, R., Gupta, R. K. and Patil, R. T. (2011). Validation of drying models and rehydration characteristics of betel (*Piper betel* L.) leaves. *Journal of Food Science and Technology,* 48(6): 685-691.

Bhakta, B., Bhakta, U. Ghanti, S., Jana, B. B., Ghanti, P., Yamane, S., Bhakta, J. N. (2016) Betelvine: A green gold for sustainable economic upliftment in rural India. *Int. J. Env. & Tech. Sc.* 3:76-82.

Das, S., Parida, R., Sandeep, I. S., Nayak, S. and Mohanty, S., (2016). Biotechnological intervention in betelvine (*Piper betle* L.): A review on recent advances and future prospects. *Asian Pacific Journal of Tropical Medicine*, 2016: 1–9. http://dx.doi.org/10.1016/j.apjtm.2016.07.029.

Madan, A., Balan, N. and Barma, R.D. (2014). Reducing postharvest losses of betel (*Piper betel* L.) leaves by various preservation techniques. *Journal of AgriSearch,* 1(4): 251-256.

Thomas, U. C., Chandini, S. and Thomas, A. (2013). Integrated nutrient management studies in betelvine *(Piper betel* L.). *International Journal of Applied Research and Studies,* 2(7): 1-9.

Chapter 14

FUTURE PROSPECTS

The betel vine is recognized as a valuable gift of nature, because, it has numerous and significant importance for human as well as environmental health points of view. It has immense social and cultural, economic, nutritional and medicinal importance as well as a folk (Siddha and Ayurvedha) reputation in each and every plethora of human life from the dawn of civilization. Therefore, leaves of the betel vine have huge demands, and are consumed by about 15 to 20 million people in India and 2 billion foreign consumers annually (Das et al., 2016). In international market, it also has significant demands and is able to earn substantial foreign remittance in betel vine cultivated countries. In India, the crop provides Rupees 6000 to 7000 million national incomes per year and at the same time leave of Rupees 30 to 40 million worth is exported to other countries (Das et al., 2016). This clearly indicates that it has a tremendous potentiality in earning the foreign exchange which can help to strengthen the nation economy in various ways. Additionally, it is a potential field for generating the huge employments and self employments (Bhakta et al., 2016).

Additionally, leaves of the betel vine possess a vast array of significant medicinal and herbal properties, for example – antioxidant, anti-inflammatory, anti-apoptotic, anti-cancer and anti-microbial, antiseptic, stimulant properties, etc. as discussed elsewhere in this book. The leaves of the betel vine contain eugenol rich 1 – 3% essential oil including chavibetol, caryophyllene and methyl eugenol essential oils as active compounds (Das et al., 2016). The cost of betel leave derived essential oil is about 10$ per 5 ml. Since, it is the potent source for medicine, stimulant, antiseptic, tonic, and other herbal and ayurvedic formulations. Such, useful properties of the oil indicate a promising industrial future of it as a raw material for manufacturing skin emollients, tooth-pastes, tooth-powders, paan masala, perfumes, room fresheners, soaps, face creams, antiseptic creams and lotions, deodorants, cold drinks, chocolates, incense sticks, appetizers, carminative mixtures digestive agents, tonics, medicines etc. (Madan et al., 2014).

It is obvious that a wide range of products from betel leaf can be manufactured at industrial level and can be sold in world markets. It can be cultivated using the approach of eco-friendly and ecological engineering technology for environmental health perspective. That is why; it is recognized as "green gold" or "green heart" of the nature for its immense human as well as environmental importance. Therefore, the betel vine has immense significant prospect in the future.

In view points of the above potentials of the betel vine, a vigorous basic research is still needed to explore and open novel and prospective avenues of its conservation, significant qualities and cultivation aspects. In this respect, betel vine research station is needed to carryout advanced and modern research by finding concerned problems discussing with the farmers directly involved in cultivation of the betel vine. In India, the research station has first established in 1989 and has started functioning from 1991 at village Diwthana Post Chandikapur in Tahsil Akot District Akola to facilitate the research work on betel vine in Vidarbha region. Moreover, the basic mandate of betel vine research should be advanced in identification of different cultivars or varieties by molecular biotechnology for conservation, exploring and identifying the important active compounds and beneficial qualities, finding of cost effective sustainable cultivation practices and protective measures in respect to this potential cash crop.

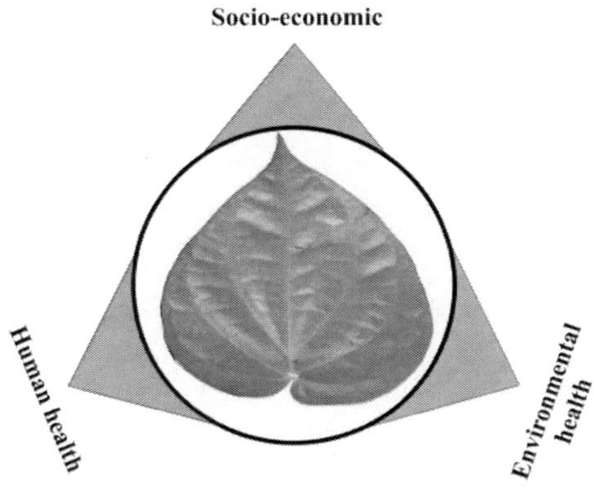

Figure 70. Potential socio-economic, human and environmental health related three-dimensional prospects of the betel vine.

On account of the various significant merits of the betel vine discussed in this book, it can be concluded that the betel vine has potential socio-economic, human and environmental health-related three-dimensional prospects in future that could play an important role in sustainable development, especially for rural peoples (Figure 70).

REFERENCES

Bhakta, B., Bhakta, U. Ghanti, S., Jana, B. B., Ghanti, P., Yamane, S., Bhakta, J. N. (2016) Betelvine: A green gold for sustainable economic upliftment in rural India. *Int. J. Env. & Tech. Sc.* 3:76-82.

Das, S., Parida, R., Sandeep, I. S., Nayak, S. and Mohanty, S., (2016). Biotechnological intervention in betelvine (Piper betle L.): A review on recent advances and future prospects. *Asian Pacific Journal of Tropical Medicine*, 2016: 1–9. http://dx.doi.org/10.1016/j.apjtm.2016.07.029.

Madan, A., Balan, N. and Barma, R. D. (2014). Reducing postharvest losses of betel (Piper betel L.) leaves by various preservation techniques. Journal of Agri Search 1(4): 251-256.

About the Author

Dr. Jatindra N. Bhakta serves as Associate Professor in the International Centre for Ecological Engineering, University of Kalyani and is associated with Faculty of Agriculture, Kochi University, Japan. He worked as research scholar, was awarded Ph. D. from University of Kalyani, India, and worked as research associates of ICAR, DBT and DST, Govt. of India. He was awarded the prestigious JSPS Postdoctoral Research Fellowship by Japan Society for Promotion of Science, Govt. of Japan and worked as Researcher Faculty at the Department of Environmental Engineering and Research Institute of Molecular Genetics, Kochi University, Japan. He has research and teaching experiences on environment, microbiology & probiotic bacteria, carbon sequestration & biofuel production, bioremediation, metagenomic, wastewater treatment & recycling, fisheries & aquaculture, agriculture and environmental pollution. Dr. Bhakta along with other collaborators (>25 researchers from 11 countries) has published more than 60 journal research articles, few books and book chapters, abstracts and popular articles. He is editorial board member and reviewer of several international journals including Elsevier and Springer Publishers. He successfully handled several projects, organized seminars and conferences, served as research mentor for Ph. D., masters and graduate students and actively associated with some NGO's especially for environmental conservation and rural socioeconomic development.

INDEX

B

C

D

E

F

G

H

I

J

K

L

M

Q

R

S

T

U

V

W